T0226974

# BASIC GAS CHROMATOGRAPHY– MASS SPECTROMETRY

## Principles and techniques

# BASIC GAS CHROMATOGRAPHY– MASS SPECTROMETRY

## Principles and techniques

**F.W. Karasek and R.E. Clement**
*University of Waterloo, Waterloo, Ontario, Canada*

**ELSEVIER**

Amsterdam – Boston – Heidelberg – London – New York – Oxford
Paris – San Diego – San Francisco – Singapore – Sydney – Tokyo

ELSEVIER SCIENCE B.V.
Sara Burgerhartstraat 25
P.O. Box 211, 1000 AE Amsterdam, The Netherlands

First edition 1988
Second impression 1991
Third impression 2003

ISBN:    0-444-42760-0

The paper used in this publication meets the requirements of ANSI/NISO Z39.48-1992 (Permanence of Paper).

Transferred to digital printing 2005

Printed and bound by Antony Rowe Ltd, Eastbourne

# PREFACE

Although this book is intended to supplement the material contained in the National Science Foundation sponsored computerized course on the subject, the SIINC (Scientific Instruction and Information Network and Curriculum), it is designed also to be a stand-alone text. The format and subject coverage is based on a course taught at the University of Waterloo over the past eight years to undergraduate applied chemistry students who alternate four-month work terms in industry with academic terms. The format, therefore, is instructional, informative and application-oriented with material presented in such a way as to be useful to a broad spectrum of people. It is not an advanced text containing only the highly developed and widely used techniques of gas chromatography–mass spectrometry (GC–MS) for use as a reference by those already familiar with the subject. Rather it presents only those concepts and material necessary to understand and use the GC–MS techniques.

The basic principles of both gas chromatography and mass spectrometry are covered to the extent necessary to understand and deal with the data generated in a GC–MS analysis. Then the focus turns to the particular requirements created by a direct combination of these two techniques into a single instrumentation system. The data generated and their use are covered in detail. The role of the computer and its specific software receives special attention, especially in the matter of compound identification via mass spectral search techniques. GC–MS–computer instrumentation has reached a plateau of excellence today so that the present commercial systems will not be obsolete for a long time to come. Therefore, a detailed description of these systems is not only informative but will be pertinent to the subject matter of this book. Finally, representative applications and results obtained with GC–MS–computer techniques are presented and chosen in such a way as to permit extrapolation of specific applications to similar problems encountered by the reader. To aid in mastering the subject matter and increase understanding, interpretation problems and suggested readings are included.

*August 1987*                                              F.W. KARASEK
                                                          R.E. CLEMENT

# CONTENTS

# CHAPTER 1

# INTRODUCTION

## 1.1. An overview

Over the past 30 years our knowledge of the chemical nature of plant and animal life and the environment of the physical world has been increased tremendously by the analytical capabilities of powerful instrumentation. The methods now in common use are so sensitive that 1 $\mu$g, an amount too small for the human eye to see, is easily detected and identified. Even one millionth of that amount, 1 pg, can still be detected and identified by some techniques. All these instrumental methods are based on relatively well known and fairly simple principles of physics and chemistry. The methods, their principles and their characteristics are summarized in Tables 1.1 and 1.2.

TABLE 1.1
INSTRUMENTAL ANALYTICAL TECHNIQUES FOR ORGANIC COMPOUNDS

| Method | Observed data |
|---|---|
| Infrared spectroscopy (IR) | Characteristic spectrum |
| Nuclear magnetic resonance (NMR) | Characteristic spectrum |
| Gas and high performance liquid chromatography (GC and HPLC) | Chromatogram of separated compounds |
| Mass spectrometry (MS) | Characteristic mass spectrum |
| Gas chromatography–mass spectrometry (GC–MS) | Chromatogram/mass spectra |

TABLE 1.2
CHARACTERISTICS OF ANALYTICAL TECHNIQUES

| Method | Sample components determined | Qualitative | Quantitative | Sensitivity (g) |
|---|---|---|---|---|
| IR | 1 | Fair | Good | $10^{-3}-10^{-6}$ |
| NMR | 1 | Good | Good | $10^{-3}$ |
| GC | 1–300 | Poor | Good | $10^{-12}$ |
| HPLC | 1–300 | Poor | Good | $10^{-9}$ |
| MS | 1–3 | Good | Good | $10^{-12}$ |
| GC–MS | 1–300 | Good | Good | $10^{-12}$ |

Most samples occur as mixtures. Even if effective sample conditioning steps are used to isolate the compounds of interest, a mixture is usually still left to be analyzed. The importance of gas and liquid chromatography is explained by their ability to separate components of a mixture. When gas chromatography is combined with mass spectrometry, a technique is created with which all components of a complex mixture of 300 compounds can be separated and identified, even though they are present in amounts as low as $10^{-12}$ g in the sample. In spite of these impressive capabilities, both techniques are based on relatively simple physical principles.

All forms of chromatography involve the partitioning of compounds between two different phases, one mobile and another stationary. Each compound in a mixture partitions to a different degree between these two phases so that as they are carried along over a bed of the stationary phase separation occurs. The longer this process is allowed to continue, the greater the separation achieved until the components emerge from the bed one by one into a detector.

Mass spectrometry is quite different. If the components separated by chromatography, or obtained in a pure form by any other method, are injected into a high vacuum where the molecules can freely move in the evacuated space, they can be broken into their constituent fragments as ions by a stream of electrons. If the ions are separated according to their mass, a definite pattern of the number of ions present at each mass will be found. This pattern, the mass spectrum, is as unique to a compound as fingerprints are to people. By the mass spectrum we can identify a compound.

Given the separating power of chromatography and the identifying power of mass spectrometry a direct combination of these two techniques gives us an incredible ability to identify all the components of complex mixtures. These mixtures can come from any aspect of human activity: the environment, food, biological tissue, industrial processes, criminal investigations and all phases of medical studies. The wealth of information pouring out of the GC–MS combination demands the use of a powerful computer dedicated to specific tasks such as extraction, storing and analysis of the essential data. This adds a new dimension and power to the analytical abilities of a GC–MS system. New techniques based solely on the computer software are being developed constantly.

The achievements of GC–MS–computer instrumentation are impressive. In 1975 such an instrument was sent on a space vehicle through 200 million miles to land on the planet Mars where it performed 14 analyses on the soil there, searching for the presence of organic life on the planet. In spite of a sensitivity of ppb no presence of organic life was found. GC–MS–computer instrumentation has made it possible for us to detect a multitude of toxic chemicals in our environment and trace their origins to the activities of our modern industrialized society. Knowledge of the sources and fates of the deadly class of chemicals known as the chlorinated dioxins is witness to these achievements. In the medical field the GC–MS technique has led to the detection and better understanding of many human metabolic disorders and diseases.

## 1.2. Computerization

Virtually every area of modern life has been touched by computer technology. In many instances this has resulted in profound changes in our daily activities. Although these changes are already taken for granted by many, it has only been in the last 10–20 years that the exponential growth of computer applications has occurred. Benefits of computerization have long been recognized by users of analytical instruments; however, only recently have the capabilities and available software for computers been at such a level that they can be used effectively by non-programmers.

In the early stages of development, information was coded onto cards and processed by large, remote computers after the analysis. These early applications consisted mainly of re-plotting data in more useful formats and calculations using the raw data that would have been tedious by other methods. The rapidly decreasing cost and size of computers, combined with increasing programming skills of chemists, expanded their range of application. Interfacing techniques made it possible for electronic signals generated by detectors of analytical instruments to be automatically converted to numbers which are input directly to computers as the signals are generated. Because computers can execute instructions in microseconds, they were used not only to acquire and analyze numbers, but also to control the operation of the instrument. More precise and rapid instrumental control made it possible to obtain greater quantities of more accurate measurements. In addition, the larger quantity of data generated made development of software techniques to analyze these data even more important. The development of computers has now made it possible to dedicate one computer to controlling an instrument, while a second computer spends all of its time obtaining and analyzing data from a detector.

Use of computers in combination with GC–MS is so important that this technique is usually referred to as GC–MS–computer, where the computer is considered an integral part of the instrumentation. The magnitude of the data handling abilities needed for GC–MS illustrates why this is so. Consider a sample containing 100 compounds that are separated by GC in a 1-h analysis. Each compound will have its typical mass spectral pattern of ionic masses and their intensities so that the analyst will have 100 of these patterns to categorize. Each mass spectral pattern or fingerprint may itself have 50 peaks, each peak being characterized by two numbers. Therefore, each mass spectral pattern may consist of 100 numbers. To analyze the data from one sample will therefore require the analyst to sort through $100 \times 100 = 10,000$ numbers. A complete analysis of the information from even a single 60-min analysis could take weeks to perform manually, but can be done in hours by a computerized system. Many applications of GC–MS–computer systems require analysis of samples having much greater complexity than that described above.

To understand and effectively apply the GC–MS–computer technique it is important to have some understanding of the role of the computer. It is not necessary to become a programming expert. The analyst should know the various types of data that can be generated and the advantages and limitations of each.

*CHAPTER 2*

# GAS CHROMATOGRAPHY

## 2.1. Fundamentals

Gas chromatography (GC) is one specific form of the more general separation process of chromatography. All forms of chromatography involve the distribution, or partitioning, of a compound between two different phases, one mobile and the other stationary. In a mixture, compounds partition to a different degree between the two phases depending upon their respective solubility in each phase. As the compounds in a mixture are carried along by the mobile phase over a fixed bed of the stationary phase they will be retarded to different extents because of their different solubility and will become separated physically. Those with greater solubility in the stationary phase take longer to emerge from the bed than those with lesser solubility. In GC the mobile phase is an inert carrier gas and the stationary phase is a high molecular weight liquid which is deposited either on the surface of finely divided particles or on the walls of a long capillary tubing. Fig. 2.1 illustrates the partitioning principle in GC.

A GC instrument, shown schematically in Fig. 2.2, has simple components. Usually helium, hydrogen or nitrogen gas compressed in cylinders is used as the carrier gas. Flow of the carrier gas into a temperature controlled sample injection device is controlled by pressure regulators and gas metering valves. A GC column is attached to the injection port and samples are introduced into the carrier gas stream at a temperature sufficient to insure vaporization of all components. Typically, the sample is introduced with a microliter syringe which is forced through a rubber septum at the injection port. A detector attached directly to the column exit monitors individual sample components as they are eluted from the column. The detector must be insensitive to the carrier gas, while detecting sample components that are eluted. A recording of its response with time forms a chromatogram. Columns are either packed with liquid-coated particles or have liquid deposited directly on the inner walls of long capillary tubing. The stationary liquid phase may be chosen from over 100 phases available, but in practice less than 10 are in common use.

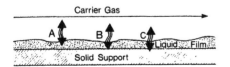

Fig. 2.1. Illustration of GC partitioning principle. A, B and C are components that partition to different degrees between gas and liquid phases as indicated by arrows.

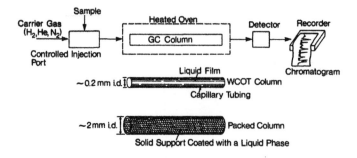

Fig. 2.2. Schematic of GC instrument.

TABLE 2.1

PARAMETERS MEASURABLE FROM A CHROMATOGRAM

| Name | Equation | Comment |
|---|---|---|
| Partition coefficient | $K_i = \dfrac{\text{conc. of "i" in liquid}}{\text{conc. of "i" in gas}}$ | Physical constant of system of "i", given column, liquid and gas and temperature |
| Capacity ratio | $k_i = K_i \dfrac{\text{volume liquid}}{\text{volume gas}}$ $k_i = \dfrac{t_i - t_m}{t_m}$ | Determines retention time of component i, and quantity of sample which can be injected |
| Theoretical plates | $N = 5.54 \left( \dfrac{t_i}{w_i} \right)^2$ | Measures column efficiency in producing narrow peaks. Value obtained depends on component |
| Resolution | $R_{ij} = 2 \dfrac{t_j - t_i}{w_j + w_i}$ | Measures column separating ability for components i and j |
| Separation factor | $r_{ij} = k_j / k_i = \alpha$ | Measures separation between i and j |
| Height equivalent to theoretical plate (HETP) | $h = \dfrac{\text{column length}}{N}$ | Measures column efficiency independent of length |
| Carrier gas average linear velocity | $\bar{\mu} = \dfrac{\text{column length}}{t_m}$ | Has optimum value for column efficiency (lowest $h$ value) |
| Effective plates | $N_{eff} = 5.54 \left( \dfrac{t_i - t_m}{w_i} \right)^2$ | Accounts for effect of column dead volume |
| Effective $h$ | $h' = \dfrac{\text{column length}}{N_{eff}}$ | Used primarily for calculations on wall-coated open tubular (WCOT) columns |
| Retention volume | $V_R = t_r F_m$ | Characteristic volume of carrier gas required to elute component r, $F_m$ is the flow rate of the carrier gas |

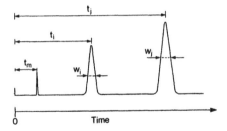

Fig. 2.3. Chromatogram and parameters measureable. $t_m$ = elution time of unretained solute; $t_i$, $t_j$ = retention times of peaks i, j; $w_i$, $w_j$ = peak width at half-height of peaks i, j.

The chromatogram contains the analytical data for the components of a mixture. Qualitative information appears in the characteristic retention time of each component. Quantitative information is contained in the peak area. A chromatogram is also a valuable measure of the performance and efficiency of the chromatographic system which produced it. Some rather simple measurements made on the chromatogram can be related to theoretical parameters of the separation process and reveal how the GC variables of temperature, carrier gas flow rate and column characteristics affect the separation. These parameters can be calculated and then used as a guide to the operation of the column and for determining the changes that can be made to achieve optimum performance for a given analysis. Understanding their theoretical basis and physical meaning is an important aspect of chromatography.

The measurements shown in Fig. 2.3 and the parameters calculated from them in Table 2.1 apply to any chromatogram. It should be emphasized that these parameters are dependent upon the specific carrier gas and liquid phase used, the quantity of liquid and gas phase present, and the physical characteristics and temperature of the column.

## 2.2. Van Deemter equation

### 2.2.1. Significance

The four chromatograms shown in Fig. 2.4 were obtained for a compound at different carrier gas flow rates. By making the measurements and calculations indicated in the figure a plot of efficiency ($h$) versus flow rate ($u$) can be constructed. Analysis of the zone dispersion phenomena occurring in a chromatographic column from a chemical engineering viewpoint led Van Deemter to propose an equation to describe the process in terms of column parameters. The development of this equation has led to many improvements in column design originating from the knowledge of how to optimize the column parameters which it provides. The equation in its simplest form contains constants $A$, $B$, $C$ which are related to physical processes occurring within the column that affect the degree with which compounds partition between stationary and mobile phases.

$$h = A + B/u + Cu$$

8

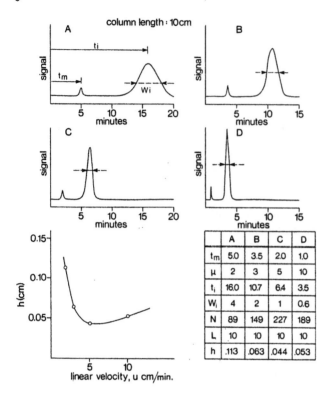

Fig. 2.4. Effect of carrier gas flow rate of GC peaks. $t_m$, $t_i$, $w_i$ as defined previously; $\mu$ = carrier gas linear velocity (ml/min); $N$ = number of theoretical plates; $L$ = column length (cm); $h$ = height equivalent to a theoretical plate (cm).

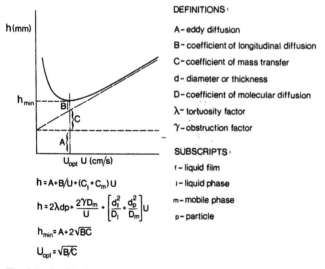

DEFINITIONS :

A – eddy diffusion
B – coefficient of longitudinal diffusion
C – coefficient of mass transfer
d – diameter or thickness
D – coefficient of molecular diffusion
$\lambda$ – tortuosity factor
$\gamma$ – obstruction factor

SUBSCRIPTS :

f – liquid film
l – liquid phase
m – mobile phase
p – particle

$$h = A + B/U + (C_l + C_m)U$$

$$h = 2\lambda dp + \frac{2\gamma D_m}{U} + \left[\frac{d_f^2}{D_l} + \frac{d_p^2}{D_m}\right]U$$

$$h_{min} = A + 2\sqrt{BC}$$

$$U_{opt} = \sqrt{B/C}$$

Fig. 2.5. Graphical representation of terms in Van Deemter equation.

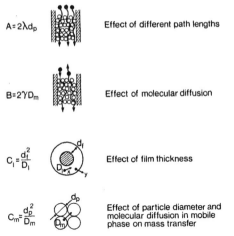

$A = 2\lambda d_p$     Effect of different path lengths

$B = 2\gamma D_m$     Effect of molecular diffusion

$C_l = \dfrac{d_f^2}{D_l}$     Effect of film thickness

$C_m = \dfrac{d_p^2}{D_m}$     Effect of particle diameter and molecular diffusion in mobile phase on mass transfer

Fig. 2.6. Pictorial illustration of physical terms in Van Deemter equation.

A more detailed description of the physical meaning of these constants is shown in Figs. 2.5 and 2.6.

A Van Deemter plot for a column reveals many things about its performance. The carrier gas flow rate giving an optimum efficiency is clearly revealed as well as how much efficiency is sacrificed as one speeds up the analysis. An examination of the various factors in the equation reveals the changes to be made in the column to achieve a desired result. Fig. 2.7 shows several plots in which the effects of changing the carrier gas and the amount of liquid phase are observed. The improved performance achieved by changing the carrier gas from nitrogen to helium and reducing the liquid phase film thickness from a loading of 30% (weight liquid to weight solid support) to 5% is clearly seen. A summary of which terms in the equation are affected by such changes is shown in Table 2.2.

The theoretically ideal column is merely an open tube on whose walls a thin film

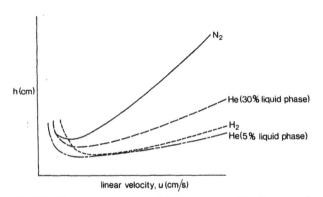

Fig. 2.7. Effect of carrier gas and liquid phase on Van Deemter plots.

TABLE 2.2
PARAMETER CHANGES AND EFFECTS ON VAN DEEMTER EQUATION CONSTANTS

| Parameter changed | A | B | $C_\ell$ | $C_m$ |
|---|---|---|---|---|
| Carrier gas: $H_2$ to $N_2$ | – | < | – | > |
| Liquid phase: 5% to 30% | – | – | > | – |
| Particle size: decrease | < | – | – | < |
| Temperature: increase | – | > | < | < |

of partitioning liquid has been deposited. These wall-coated open tubular (WCOT) columns are commonly known as capillary columns since their inner diameters (I.D.) are so small (0.2 mm). The theoretically derived equation for WCOT columns has the same form as the Van Deemter equation and allows us to analyse their operation as can be done for packed columns.

$$h = \frac{2D_m}{\mu} + \left[\frac{1 + 6k + 11k^2}{24(1 + k)^2}\right]\left[\frac{r^2}{D_m}\right]\mu + \left[\frac{k}{6(1 + k)^2}\right]\left[\frac{d_f}{D_f}\right]\mu$$

Since the equation (see Figs. 2.5 and 2.6 for definition of terms) shows that $h$ is directly proportional to $r^2$, the limit on efficiency achievable by reducing the radius becomes a practical one of producing workable instrumental components for such narrow columns, especially for sample injection. Standard WCOT columns are in the 0.2–0.3 mm I.D. range.

### 2.2.2. Evaluation

When the Van Deemter equation first appeared in 1956 it served an important function in unifying our concepts about chromatography. In the intervening years numerous attempts to refine the equation theoretically have deepened our understanding of the fundamental processes involved. We can now take a critical look at the equation and what it explains.

The most rigorously correct equation current theory allows is shown in Table 2.3 along with definitions of the terms involved. The terms in this equation can now be examined individually.

*The A term: multiple path or eddy diffusion.* Van Deemter conceived this term to describe processes occurring in pockets between particles. The oversimplified picture of molecules taking long or short paths resulting in band broadening soon had to be abandoned because as particles became smaller in size and more uniform, careful experiments revealed no measureable A term. The concept of an A term may be correct, but it must be coupled to diffusion and flow velocity. We can understand this better by considering what happens to a molecule first near zero, and then near infinite, flow velocity. Near zero velocity the molecule will move from one path to another many times by diffusion while traversing the column and no molecule will follow an individual short or long path. Near infinite velocity the molecule would be

TABLE 2.3
CURRENT REPRESENTATION OF THE VAN DEEMTER EQUATION

$$h = \frac{2_\gamma D_m}{\mu} + q\frac{k}{(1+k)^2}\frac{d_f^2}{D_\ell}\mu + \frac{f_n(d_p^2, d_c^2, \text{end}, \mu)}{D_m}\mu$$

*Definitions*
$q$ = factor describing shape of liquid phase
$c$ = subscript referring to column
end = end-effects
Other symbols as indicated in Figs. 2.5 and 2.6

locked into an individual path, whether it be short or long. This is illustrated in Fig. 2.8. There is an equation which presumes to describe this:

$$h = \frac{1}{1/A + 1/(C_m u)}$$

This equation seems to describe the $A$ term at zero and infinite values of $u$, but much experimental evidence shows it does not do so at other values of $u$. The equation is based on laminar flow, whereas at high velocities turbulence becomes important. There appears to be no equation giving a quantitative treatment of this term, so it seems best to drop it for packed columns and incorporate the effect in the $C_m$ term.

*The B term: longitudinal diffusion.* This term as originally expressed by Van Deemter, $2VD_m/u$, appears to be essentially correct. Evidence, both theoretical and experimental, shows that the obstruction factor $V$ decreases at low velocities, but the expression does describe adequately the phenomena occurring in the column.

*The $C_\ell$ term: mass transfer in the liquid phase.* The equation Van Deemter used to express this process was:

$$C_\ell = (8/\pi^2)\left[k/(1+k)^2\right]\left(d_f^2/D_\ell\right)$$

The $8/\pi^2$ appears to be incorrect because Van Deemter made some wrong assump-

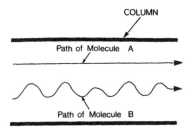

Fig. 2.8. Illustration of different paths possible for molecules travelling through a GC column.

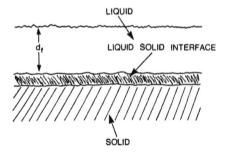

LIQUID

LIQUID SOLID INTERFACE

$d_l$

SOLID

Fig. 2.9. Illustration of forces affecting mass transfer in liquid phase.

tions. Others have theoretically arrived at a value of 2/3 for this constant, but the assumptions made, such as perfectly uniform film, are not true in practice so it is best to use a shape factor "$q$" instead as a more realistic constant. The shape factor is related to the shape of the particles and the size and depth of pores which hold the liquid phase. One can visualize these occurring in such a way as to give $q$ values between 2/15 and 2/3.

There is one further complication in this picture of the liquid phase. The assumption is made that the support upon which the liquid is deposited is essentially inert. For the liquid to adhere, however, the support must be adsorptive. Therefore, molecules diffusing through the liquid phase must face still another force which reduces their mass transfer through the phase. Some will become adsorbed at the liquid-support interface and hence spend a longer time than the average in the liquid phase (Fig. 2.9). There are equations to express this, but the factors in them are difficult to evaluate. A qualitative evaluation of these factors leads to the important concept that an adsorptive support surface can have a serious effect on efficiency.

$C_m$: *mobile phase mass transfer.* The complexities of mass transfer in the mobile phase are too great to permit an exact analysis for packed columns. However, the causes of zone broadening are qualitatively understood well enough to lead to dramatic improvements in column performance and are expressed as some function of column parameters by:

$$C_m \propto \frac{d_p^2, \ d_c^2, \ \text{end}, \ u}{D_m}$$

These individual functions indicate that $h$ increases with $d_p^2$, $d_c^2$, $d_c$/coil diameter, geometry of the column and decreases with uniformity of packing and efficiency of the end connectors of the column.

*Concept of particle diameter, $d_p$.* The importance of particle diameter arises from the effect of the stagnant mobile phase within the pores of the particles and the flow pattern between particles. As depicted in Fig. 2.10, molecules in pores lag behind those in the flowing stream, so that equilibrium cannot be achieved in practice.

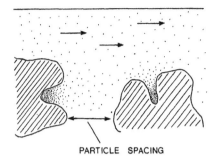

PARTICLE SPACING

Fig. 2.10. Effect of pore geometry of mass transfer.

Molecules in the pores lag behind for a time that depends on the distance they must diffuse to leave the particle and associated liquid phase. This distance is proportional to the particle diameter. Also, the particle must be reached by diffusion from the flow stream, which is proportional to the diameter of the particles, and this leads again to $C_m$ being proportional to $d_p^2$. These effects can be written as:

$$C_m = \gamma d_p^2 / D_m$$

The ill-defined constant, $\gamma$, contains the geometrical effects introduced by the packing structure which can vary across the diameter of the column with particle sizes. The small particle sizes used in high performance liquid chromatography (HPLC) are less affected by this factor.

*Concept of column diameter, $d_c$.* A theoretical equation can be derived for open tubular columns which is rigorously correct and consistent with experimental data.

$$C_m = \frac{(1 + 6k + 11k^2)d_c^2}{96(1 + k)^2 D_m}$$

Since the model for this equation assumes a parabolic flow profile and smooth walls, it would fail when the column diameter reaches 0.0125 cm or lower.

There are effects on $C_m$ attributed to the diameter of the coiling of the column because the flow path on the inside of the coil is shorter than on the outside. In practical GC columns these are small, indicating qualitatively that tighter coils improve efficiency, and in the straight HPLC columns used they do not exist.

*Concept of end-effects.* Zone broadening occurs in the injection system and in the connections of the column to injector and detector. The dead spaces in these connectors and the uneven flow profile caused by their geometry gives significant contributions to the $C_m$ term (Fig. 2.11). The longer the column the less important is this contribution, which makes it inversely proportional to column length. A practical consequence of this is that connecting tubes should be as narrow as possible, or be packed with inert glass beads of the same diameter as the column packing.

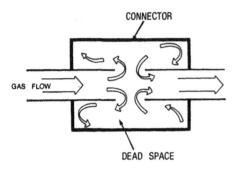

CONNECTOR

GAS FLOW

DEAD SPACE

Fig. 2.11. Effect of dead space on $C_m$ term.

In GC the sample is often evaporated in the injection system. A finite rate of evaporation gives an exponential spread to the zone. This spread increases with carrier gas velocity but it is not reduced by diffusion so that is part of the $C_m$ term, but does contribute to the $C_s$ term.

*Use of the equation.* Because of the approximations involved in its derivation, any form of the Van Deemter equation is strictly applicable only to infinitely long columns. The approximations become more important as the plate number, $N$, becomes smaller. It is certainly inapplicable for values of $N$ below 100, and doubtful below 1000. It also appears to be inappropriate to apply the equation to columns that are less than 50 times as long as their diameter. These limitations do not affect the qualitative usefulness of the equation, but do make one aware of its area of applicability.

## 2.3. Columns

GC columns most widely used fall into two distinct categories: packed and WCOT. Packed columns were developed first but since about 1980 the commercial availability of highly efficient and rugged fused silica WCOT columns has resulted in their use dominating the GC–MS field, especially for the analysis of trace amounts of specific organic compounds in complex mixtures. Despite the current popularity of WCOT columns, it is important to understand the advantages and limitations of both packed and WCOT columns for GC–MS analysis.

In all GC columns there are two important characteristics which must be achieved. The stationary phase must be created in a controlled manner with a film of uniform thickness completely covering the support surface which is as inert as possible. The film must be thermally stable both to decomposition and vaporization up to the highest temperatures necessary to conduct the analysis. This will usually involve chemically bonding stationary phase molecules to the support surface molecules or creating a polymer-like film held in place by physical forces. Having an inert surface requires elimination of all active sites on the support. Active sites are created primarily by exposed areas of the support whose properties cause irreversible adsorption of sample components and whose metallic elements act as catalytic

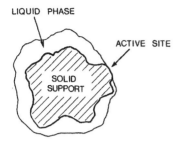

LIQUID PHASE

ACTIVE SITE

SOLID
SUPPORT

Fig. 2.12. Representation of surface of GC support material for packed column showing active site.

sites for decomposition of specific components (Fig. 2.12). The effect of these active sites becomes more and more pronounced as the quantity of the sample components decreases. Columns whose active sites cause little effect at the microgram range may fail drastically for compounds at the picogram range.

Fig. 2.12 shows how active sites are the result of incomplete coating of solid support particles in packed column GC. It is also shown that the liquid phase coating is not evenly distributed around the particle: some locations are more thickly coated than others. Consequently, if the column is operated at or beyond its upper temperature limit for extended periods, then more liquid phase will be bled off of the particle surface and additional active sites will be formed. Eventually, the column will have to be discarded. Although WCOT columns do not have solid support particles, active sites can also be significant if a smooth, even coating of liquid phase on the walls of the column is not achieved.

### 2.3.1. Packed columns

Packed columns do not possess the high efficiencies and separating capabilities of WCOT columns, but they do have higher capacities which simplifies sample introduction techniques and provides a larger quantity of sample component for introduction to the detector. Typical packed columns for analytical work are 2–5 m in length and have 2 mm I.D. They use carrier gas flows of 20–50 ml/min. Larger columns can be designed specifically for preparatory-scale GC.

In preparing these columns all the considerations detailed in the Van Deemter equation must be observed: use uniform, inert particles of small size for the support; prepare an even, thin coating of a thermally stable stationary phase; and pack the columns to ensure close, uniform spacing of the particles. Care must be taken to avoid fracture of the particles which exposes adsorptive surfaces and active sites. The most common support particles are formed from diatomites, which are skeletons of a single cell algae. These are prepared by molding diatomaceous earth into bricks and drying in an oven. The bricks are then crushed and screened to particles in the 80–120 mesh size (80–120 openings per inch in a screen).

These particles are very strong with a high specific surface area (1–20 $m^2$/g) and good pore structure (1–2 $\mu$m). They are poor for use with polar stationary phases. Supports which are much better for polar compounds can be are formed by adding

2% sodium carbonate to diatomaceous earth and heating to 900°C. This procedure fixes individual diatoms into larger, but more fragile particles. All these supports have Si–OH groups on the surface which must be deactivated by chemically bonded reactants, and about 10% metallic impurities such as iron and aluminum that must be removed by acid washing.

Use of high concentrations of stationary phase in the 20% range (weight percent of solid support) has the advantage of more complete coverage of any active surface area present and provides a high sample capacity. Such columns have lower efficiencies and produce bleed under temperature programming, making them unsuitable for GC–MS application. Stationary phase concentrations below 3% are more suitable. These columns have many advantages: higher efficiency, less bleed, easier to pack the column uniformly, more rapid analysis and component retention times which occur at lower temperatures. High efficiency packed columns have been developed with 0.2% stationary phase that are especially suited for GC–MS application. These are prepared by coating a 6% Carbowax-20M liquid phase on a Chromosorb support of which the surface has been cleaned and the metallic elements have been removed by exhaustive washing with hydrochloric acid. The coated particles are then thermally treated in a nitrogen atmosphere at 200°C overnight, after which the liquid phase is removed by exhaustive washing with methanol. The remaining 0.2% polymer film of Carbowax completely covers the particle surface in an even layer. A column made of this packing material shows high efficiency, low bleed and almost no active sites and functions well in GC–MS applications. Because of the thin layer of liquid phase, such columns are also more susceptible to decomposition by oxygen and water when operated at their upper temperature limit (about 250°C). This can be easily prevented by installing a special sorbent cartridge to trap these contaminants between the column and carrier gas supply.

### 2.3.2. Wall-coated open tubular columns

WCOT columns were originated in 1956 when Golay first showed them to be theoretically ideal. The early WCOT columns were made of metal, but the problems of eliminating their active sites and those of producing acceptable films of the stationary phase on metal surfaces resulted in columns of poor performance and retarded their acceptance. WCOT columns received major advances when pioneering work on high performance glass WCOT columns occurred in Europe in the period 1975–1980. Their major disadvantages were those of fragility and the need for use of skillful technology in deactivation and coating techniques.

The development and widespread availability of fused silica WCOT columns has changed the picture drastically. These columns are made of pure $SiO_2$ and are extremely rugged when an external coating of polyimide polymer is applied. The technology of producing high quality columns of controlled internal diameter and stationary film thickness has advanced considerably. Commercially available fused silica WCOT columns with both polar and nonpolar stationary phases have a consistently high efficiency and give excellent analytical results. They are clearly the columns of choice for GC–MS application. Fig. 2.13 illustrates the essential

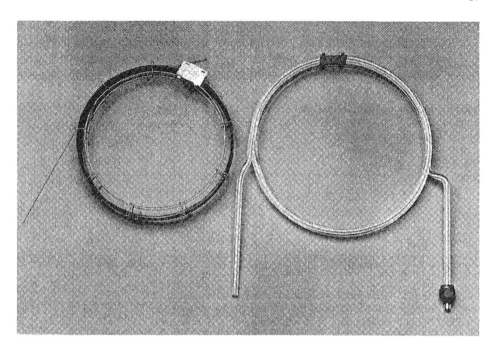

Fig. 2.13. Comparison of 2-m packed column (right) and 30-m fused silica WCOT column (left). The 80–100 mesh coated particles of the packed column are clearly visible through the glass.

configurations of these columns. They are produced in lengths of 10–100 m with I.D.s of 0.20–0.35 mm and use carrier gas flow rates of 2–5 ml/min. This permits the use of GC–MS interfaces that transmit the maximum amount of organic material to the mass spectrometer ion source. Fused silica WCOT columns have the additional advantage over glass or metal columns of being physically flexible, so much that they can actually be tied into a knot without breaking. Therefore, fused silica WCOT columns can be installed in every model of GC–MS, regardless of the configuration of the hardware. Glass columns must be custom fit for each configuration by heating the end of the column until it can be formed. Such thermal treatment can destroy the liquid phase in a portion of the column.

Table 2.4 compares the important characteristics of packed and WCOT columns. WCOT columns have stringent sample introduction requirements. Along with the development of the fused silica WCOT column has come the perfection of sample introduction systems to meet these requirements.

## 2.4. Detectors

In gas chromatography the detection problem is that of sensing a small quantity of organic compound in an inert carrier gas. These quantities can vary from $10^{-3}$ g in 10 ml of gas to $10^{-12}$ g in 0.05 ml. To detect these compounds under such a wide

TABLE 2.4

ESSENTIAL CHARACTERISTICS OF GC COLUMNS

| Parameter | Packed | WCOT |
|---|---|---|
| Length (m) | 1–6 | 10–100 |
| I.D. (mm) | 2–4 | 0.20–0.35 |
| Total plates ($N_{eff}$) (2 m packed; 50 m WCOT) | 5000 | 150,000 |
| Capacity | 10 $\mu$g/peak | 50 ng/peak |
| Thickness stationary phase ($\mu$m) | 1–10 | 0.05–0.5 |
| Carrier gas flow rate (ml/min) | 10–60 | 0.5–10 |
| Column pressure drop (p.s.i.) | 10–40 | 3–40 |

TABLE 2.5

COMPARISON OF CHARACTERISTICS OF GC DETECTORS

| Detector | Principle | Selectivity | Sensitivity (g/s) | Linear range |
|---|---|---|---|---|
| Thermal conductivity detector (TCD) | Measures difference in thermal conductivity of gases | Universally responds to all compounds | $10^{-10}$ | $10^4$ |
| Flame ionization detector (FID) | Burns compounds in $H_2/O_2$ flame at 2000°C and measures ions created | Responds to all oxidizable carbon compounds | $10^{-12}$ | $10^7$ |
| Electron-capture detector (ECD) | Measures changes in electron current caused by reaction of organic compound with electrons | Responds to all electron reacting compounds | $10^{-14}$ | $10^3$ ($10^6$ pulsed operation) |

variety of conditions a number of detectors have been developed. The ideal detector will have high sensitivity, wide linear dynamic range and a small cell volume so the GC peak is not distorted. Those in most common use are the thermal conductivity, flame ionization and electron-capture detectors. Their characteristics are outlined in Table 2.5.

## 2.4.1. Thermal conductivity detector

The thermal conductivity detector (TCD) was one of the earliest GC detectors developed. It is based on the simple concept that there is a difference in the thermal conductivity of the carrier gas when the organic compound is present in the GC effluent peak. The TCD is a relatively simple and sensitive device. It depends on measuring differences in the transfer of heat from its hot sensing element through the column effluent gas to the walls of the TCD cell. Design of the TCD must attempt to render it insensitive to the operating variables of the GC column and to

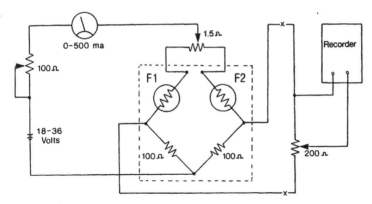

Fig. 2.14. Principle of the thermal conductivity detector. GC peaks are generated by different thermal conductivities of gases flowing across hot filaments F1 and F2.

room temperature, carrier gas flow rate and ambient pressure.

The basic principle of the TCD is illustrated in Fig. 2.14. The circuitry is that of the well known Wheatstone bridge. The resistors shown are heated by the current passing through them and they are constructed of material whose resistance changes greatly with temperature. At balance all the four resistors are equal and the output voltage at the centre of the bridge is zero. The resistance of a filament $F_1$ is determined by the thermal conductivity of the column effluent, and the resistance of filament $F_2$ is determined by that of the carrier gas. As long as no GC peak emerges these two filaments are surrounded by only carrier gas and their resistance values will be equal. When a GC peak emerges and passes $F_1$, the filaments resistance value will change because the different thermal conductivity of the GC peak causes its temperature to change. The unbalance in the bridge appears as a voltage across the centre which will reproduce the contour of the GC peak. The actual design of the TCD is more sophisticated than this diagram implies, but it is nevertheless a simple, rugged, non-destructive device capable of measuring $10^{-9}$ g/ml at its ultimate sensitivity.

An ingenious design of the TCD which eliminates many of the problems inherent in the principle is shown in Fig. 2.15. The GC column effluent and the pure carrier gas are alternately passed through a single filament detection cell. By switching back and forth at 10 cycles per second a modulated intermediate signal is obtained with an amplitude proportional to the differences between the thermal conductivity of the column effluent and carrier gas. Since the detector is only sensitive to changes occurring at 10 cycles per second, and very few changes in the operating variables can occur that fast, it only responds to the GC peak. The modular valve shown is a fluidic device having no moving parts. Its operation causes the changing flows shown by the differential pressures it creates across the filament channel.

### 2.4.2. Flame ionization detector

The flame ionization detector (FID) has all the characteristics of the ideal

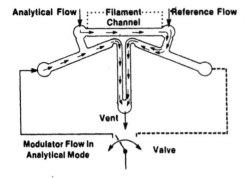

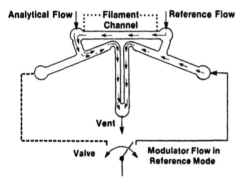

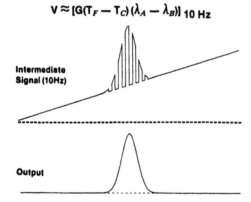

$$V \approx [G(T_F - T_C)(\lambda_A - \lambda_B)] \; 10 \; Hz$$

Fig. 2.15. Design of the Hewlett-Packard TCD detector.

detector for GC. It is based on the simple principle of combusting the GC effluent in a hydrogen flame and measuring the ions formed from the GC peak component. The FID has a minimum detectable quantity (MDQ) of $10^{-11}$ g and a linear dynamic range of $10^7$. The response is extremely fast, the detector cell volume is small and it is a simple, rugged device which is relatively insensitive to fluctuation in

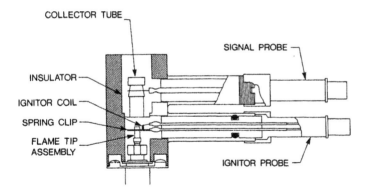

Fig. 2.16. Principle of the flame ionization detector.

most of the operating variables of carrier gas flow rate, room and column temperature and hydrogen/oxygen flow rates.

The basic elements of the FID are shown in Fig. 2.16. A flame is created by burning hydrogen in air or oxygen and the negative ions formed by the combustion are measured by applying a positive voltage of 200–300 V to a collector electrode. The hydrogen flame produces very few ions (about $10^{-14}$ A current), but when an organic compound is present in a GC peak large quantities of ions proportional to the number of organic molecules are created. Ion currents as high as $10^{-6}$ A are observed. A recording of the collected ion current versus time will reproduce the concentration of organic compound in the GC peak. Since hydrogen flow rates of 30 ml/min and air flow rates of 400 ml/min are used, introduction of the 1–15 ml/min flow of the GC column effluent into that stream results in it being swept rapidly through the detector giving fast response times. Response speed is especially important for use with WCOT columns whose GC peak widths occur in seconds. A number of compounds will not be detected: hydrogen, nitrogen, water, hydrogen sulfide, sulfur dioxide, ammonia and carbon dioxide. Carbon compounds detectable must be capable of undergoing oxidation. The FID response depends upon the numbers of ions produced for a compound. Since this varies considerably between compound classes, FID response factors will vary correspondingly.

### 2.4.3. Electron capture detector

From the time of its discovery in 1960 the electron capture detector (ECD) has enjoyed a steady growth in development and use. The wealth of interesting theoretical concepts for the electron and ion-molecule reactions involved has led to many studies to understand the basic phenomena. In spite of early problems of anomalous responses and erratic operation, the simplicity of a device giving such high sensitivity and selectivity to important classes of compounds, such as halogenated pesticides, has encouraged development of detectors with ever-increasing scope and reliability.

*Direct current ECD.* Fig. 2.17 illustrates the basic elements and functions of an ECD. High-energy electrons from a radioactive beta emitter create copious numbers

22

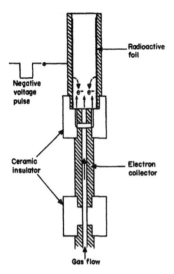

Fig. 2.17. Principle of the electron-capture detector. Each electron from a $^{63}$Ni radioactive source creates positive ions and low-energy electrons in the GC carrier gas to form a standing current ($I$). An electron-reacting substance AB will reduce the standing current to reproduce the GC peak. A plot of $I$ versus concentration of AB shows a limited linear dynamic range.

of low-energy electrons and positive ions in the nitrogen GC carrier gas. It is the reduction in collection of these low-energy electrons at the anode of the ECD, caused by their reactions with a GC peak compound, that gives an electrical reproduction of the GC peak. Although the electron current peak is a negative one, it can easily be reversed electronically. The detector is designed and operated to collect only the electrons and to avoid collecting the negative ions formed, primarily by promoting their recombination with the positive ions present. Kinetics favour this. The positive-ion concentration is several thousand times greater than the free-electron concentration, reaction rates for recombination of positive and negative ions are $10^5$–$10^8$ times faster than for positive ions and electrons, and the mobilities of the heavier ions are much lower than those of the electrons. Use of electron concentration as a measurement gives the detector its specificity. Only those compounds that react with electrons are detected. Sensitivities vary by a factor of $10^4$ between strong and weak electron reacting compounds.

The equation which describes the response of the ECD is

$$\frac{I_s - I}{I} = kC$$

where $I_s$ is the initial standing current, $I$ is the current present after reaction with a sample of concentration $C$ and $k$ is a proportionality constant. This type of response is essentially a non-linear one and results in a limited dynamic range of $10^3$.

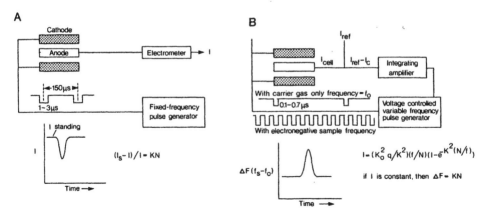

Fig. 2.18. Design of the pulsed electron capture detector. Two modes of operation are shown, fixed frequency (A) and variable frequency (B). $I$ = electron current collected; $I_s$ = electron current with only carrier gas; $K_1$, $K_0$ = cell parameter constants; $f$ = pulse frequency; $N$ = sample quantity.

The response of the detector is dependent upon extracting unreacted, low-energy electrons from a complicated plasma using a simple pair of electrodes polarized with a constant dc voltage. Detector behaviour will necessarily be dependent upon almost all the associated parameters: the dc voltage, contact potentials caused by adsorption of sample components on electrode surfaces, space charges of the slow-moving ions surrounding the electrodes, cell and electrode configuration, trace impurities, flow rate and type of carrier gas, and the temperature. Because of the unpredictable nature of these parameters, a dc-operated ECD will exhibit anomalous responses, drifting baseline, sensitivity variations and a limited, variable dynamic range.

*Pulsed mode ECD.* A substantial improvement over dc operation is achieved by applying the detector voltage as a sequence of narrow pulses with a fixed duration and amplitude (1–3 μs, 50 V) sufficient to collect the very mobile electrons but not the heavier, slower ions (Fig. 2.18). Sufficient radioactive material is used to generate about $10^{-9}$ A of thermal electrons in a suitable carrier gas, preferably a mixture of argon with 10% methane. The argon–methane mixture not only gives high mobility to the electrons but quickly moderates their energy, through collision, to a thermal level where they can more easily react with GC peak components. This mode of operation greatly minimizes many of the undesirable effects of dc operation, particularly those of space charge, unwanted ion collection and contact potential. Measuring the electron current produced by fixed-frequency pulses means the current still varies over a limited linearity range, as in the dc operation. By manually selecting the pulse frequency, different concentration regions of linearity can be chosen to minimize the linearity problem.

Fig. 2.18 also illustrates another mode of pulsed operation for an ECD in which the frequency of the pulse is automatically varied to maintain the collected current constant when a sample component enters the detector. The electron current is only collected when a pulse of positive voltage is applied to the collector anode. This current is compared to a reference current and the electronics are designed to maintain the difference between the cell current and reference current at zero. When an electron reacting compound enters the cell, the pulse generator must create many more pulses in order to generate enough electron current to have it equal the reference current. If the frequency change is then plotted versus time, the GC peak will be reproduced. The important consequence of this technique is that the frequency is directly proportional to concentration ($F = kC$) and a range of linearity of $10^6$ is achieved. Most constant-current ECD units operate with a baseline frequency of a few hundred hertz which increases to as high as 200 kHz with the sample present.

*Other factors in ECD response.* Collection of unreacted electrons is the only means of sensing the presence of the compound being detected in an ECD, however, many factors other than compound concentration influence the numbers of electrons present. These factors include trace components entering the system, temperature, cleanliness of the detector surfaces and stability of the electron source, the beta emitter. The most critical of these is cleanliness of the carrier gas and the detector elements. The effects of trace levels of oxygen, even those below 10 ppm, are not always appreciated. Data indicate that the standing current can be reduced to less than half its maximum value by the presence of 10 ppm oxygen. Although the oxygen removes electrons by formation of $O_2^-$ ions the further reaction of these $O_2^-$ species with the sample compound also adversely affects analytical results. In addition, these effects are strongly temperature-sensitive. With a fixed amount of oxygen, the relative amount of electrons at equilibrium increases with temperature, again leading to uncertain results.

Oxygen and similarly acting compounds can enter the system through connectors and any device (valves, regulators, tubing) in the instrument system, or from bleed and decomposition of the column liquid phase. Such contaminants can be reduced to acceptable levels by the use of special sorbent traps for the GC carrier gas, high purity gases for the detector cell, and by careful column installation to minimize leaks.

## 2.5. Analysis techniques

The single mandatory requirement of a sample to be analyzed by GC is that it be in the vapour state. This includes gaseous, liquid, and in some cases, solid samples. Often a solid can be dissolved in a suitable solvent and be converted to a vapour by temperature of the inlet upon injection. Some solids that cannot be dissolved in a suitable solvent can still be analyzed after conversion to a more volatile substance by chemical reaction. Certain types of non-volatile solids can also be analyzed after undergoing pyrolysis. Suitable solvents are those which will not physically harm the GC column and are amenable to the method of detection employed. For example, a chlorinated hydrocarbon would not be used as a solvent with an ECD, and water would usually not be used if mass spectrometric detection is employed. Some solvents cannot be used because they are not compatible with the specific GC column employed. Use of such solvents would result in excessive tailing that will mask early-eluting peaks.

### 2.5.1. Sample injection

Sample injection in a GC analysis is critical. Poor injection technique can reduce column resolution and the quality of quantitative results. The sample must be injected as a narrow band onto the head of the column and contain a composition truly representative of the original mixture. The amount of sample injected must match the capacity of the column to prevent overloading the stationary phase, causing degradation of column performance.

Several methods have been developed to introduce samples onto the GC column. Gases can be injected using a rotary valve containing a sample loop of known volume. Some samples are collected on a sorbent material, from which they are introduced to the GC by thermal desorption. The most common technique is liquid injection through a self-sealing septum into a heated injection port. Design of the injection port is critical to the success of the GC analysis, especially when WCOT columns are employed. For quantitative work, sample injection must be achieved with a high degree of precision.

Fig. 2.19 shows a typical design of an injection system for packed columns. The injection region is held at a high temperature, usually 200–300°C, so that the injection liquid undergoes a process termed flash vaporization. This transforms the solvent and associated compounds comprising the sample into the vapour state in a rapid and efficient manner. Carrier gas flows into the injection volume and sweeps the vaporized sample through the column. The seal between the column and injection port liner must be leak-free so all of the sample reaches the column. Empty volume or dead space around the column in the injection area should be minimized, but is not critical if high carrier gas flows are employed. Ideally, all of the sample should be rapidly swept as a small plug onto the front of the column. If this is not

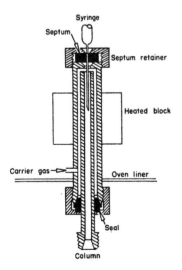

Fig. 2.19. Typical flash vaporization sample injector for packed columns.

achieved decreased column efficiency will result. Best results are obtained by ensuring rapid and complete sample vaporization, eliminating gas leaks in the injection region, and operating at high carrier gas flow rates.

*WCOT injection modes.* Capillary or WCOT columns require more sophisticated injector designs than do packed columns. Since column flow rates are only one-tenth of those used in packed columns, excessive dead volume will cause serious peak broadening. Contamination on the walls of the injection port liner can also affect results by adsorbing sample components or introducing artifacts. Contamination can also result from trace impurities in the septum that bleed onto the column due to the elevated injection port temperature. For this reason, most WCOT injector designs have incorporated a small carrier gas flow across the surface of the septum to purge impurities from the injector.

Injector designs are so critical for WCOT columns because, in addition to the low

TABLE 2.6

COMPARISON OF WCOT INJECTION TECHNIQUES

|  | Split | Splitless | On-column |
|---|---|---|---|
| Application | Principal components only | Trace and principal components | Trace and principal components |
| Maximum concentration | Depends on split ratio | 50 ng/component | 100 ng/component |
| Precision | Poor | Good | Excellent |
| Injection temperature (°C) | 250–320 | 200–280 | Injector not heated |
| Initial column temperature | Variable | 20–40°C below solvent boiling point | Near solvent boiling point |
| Advantages | Control of split ratio prevents column overloading | Direct quantitation Better than split for trace analysis Good precision compared to split | "Cool" injection reduces sample discrimination Thermally unstable compounds can be analysed Excellent precision and accuracy Direct quantitation |
| Disadvantages | Possible sample discrimination during split Flash vaporatization Indirect quantitation Poor for trace analysis | Flash vaporization Limited number of solvents compatible with solvent effect Sample must be reconcentrated by solvent effect or cold trapping | Non-volatile components will accumulate at head of column Some solvents could damage certain columns Autosamplers still in early stages of development |

operating flow rates employed, the sample capacity of these columns is very low. While as many as 10 μl of solvent containing a total of 100–500 μg of sample components may be injected onto packed columns, WCOT injection volumes are typically no greater than 2 μl, containing a maximum of 1–10 μg of total sample components. Although injector design depends upon the factors already described for packed columns, different modes of operation have been developed for WCOT columns. General principles of these injections techniques are summarized in Table 2.6.

*Split injection.* Most early WCOT inlet systems used this type of sample introduction. It is a method whereby the injected sample can contain components at concentrations similar to those normally used in packed column work. To prevent overloading of the column the sample, after undergoing flash vaporization in the heated injection region, is split into two unequal portions. The larger portion is vented from the system while the smaller is carried onto the column. Size of the split may be 10:1–50:1 for large-bore WCOT columns or 50:1–500:1 for narrow-bore columns, and is controlled by a variable valve attached to a vent line leading away

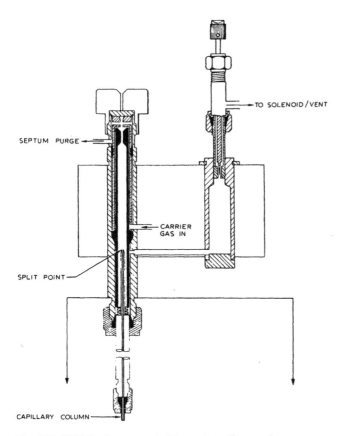

Fig. 2.20. WCOT column sample injector for split operation.

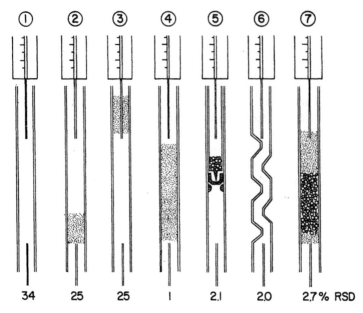

Fig. 2.21. Vaporization tubes for homogenization of sample components during split sampling. RSD values for each are given.

from the injection region. Fig. 2.20 shows one design for an injector that is operated in the split injection mode.

Although the split method of injection does prevent column overloading the fraction which reaches the column may not be representative of the original sample. Because it is a flash vaporization technique, higher molecular-weight (low volatility) components of the sample in contact with the metal surface of the syringe plunger are not expelled from the syringe with the same efficiency as compounds whose boiling points are at or below the injection temperature. Since low injection volumes are generally used for WCOT columns as compared to packed columns, a significant fraction of the sample is in contact with the metal surface and split discrimination may result. To prevent this, inlet liners such as illustrated by the seven different types of liners in Fig. 2.21 are employed. They are designed to provide efficient heat transfer and thorough mixing to the sample to minimize discrimination. In addition, the packing material in the insert prevents non-volatile material from reaching the column and causing a decrease in column efficiency.

*Splitless injection.* The split technique just described is most beneficial for samples containing compounds at high concentrations. Components present at trace levels may be undetected since most of the sample is vented and does not reach the detector. For analysis of trace compounds the splitless injection technique is generally used. For narrow-bore columns, sample components should be no greater than about 100 ng per component, and preferably 50 ng or less, to avoid column overloading. Wide-bore columns can accomodate about 100–200 ng per component, depending on the number of compounds present. The sample undergoes flash vaporization and is carried onto the column where the sample is reconcentrated at the head of the column. Without reconcentration, the volume of the injection region will increase the band widths of eluting peaks and reduce the efficiency of the separation.

One method of reconcentration is known as the solvent effect. In simple terms, this occurs because the front of the solvent plug which enters the column mixes with the stationary phase and is more strongly retained than the rear of the solvent plug. Therefore, sample components encounter a barrier which has the effect of condensing components at the head of the column. Because of the interaction of solvent and stationary phase, some columns can be damaged if polar or aromatic solvents are used. Common solvents employed are methylene chloride, chloroform, carbon disulfide, dimethyl ether, hexane and isooctane. Some of the new cross-linked fused silica columns have immobilized phases which will not be damaged even by aromatic or polar solvents. To accomplish this solvent effect, it is necessary that the column temperature at injection is low enough to prevent the solvent from migrating too rapidly from the head of the column. This requires a column temperature of 20–40°C below the boiling point of the solvent, and may require auxiliary cooling of the oven.

A second means of solute reconcentration is cold-trapping. By this method, the initial column temperature must be about 150°C below the boiling points of the components to be trapped. Compounds with lower boiling points require a solvent effect for reconcentration.

Fig. 2.22 shows one method of performing splitless injection. At injection, the column carrier gas flow rate is the same as the inlet flow rate and sample components are swept on to the head of the column where they are reconcentrated by cold trapping or the solvent effect (Fig. 2.22A). If left under these conditions the time to sweep all off the solvent vapour onto the column would be great enough to cause a large solvent tail which would interfere with early-eluting peaks. Therefore, after a specified time (about 30 s) the operation of a solenoid valve causes conditions to change so that the inlet flow is greater than the column flow, which remains constant (Fig. 2.22B). Solvent that still remains in the inlet will be purged, resulting in a much sharper solvent peak. Note that the same inlet system illustrated in Fig. 2.22 can be used for different modes of operation by proper choice of inlet liner and operating conditions.

*Sandwich sampling techniques.* It is important to inject the entire sample in a discrete plug into whatever type of injector is used. Evaporation from the needle point and leaving residual traces of the sample on the interior of the needle must be avoided. The most effective way to do this is to use a technique in which the sample is placed between two segments of air followed by a segment of solvent. When the plunger of the syringe forces this sandwich out the air cushion maintains integrity of the sample and the segment of solvent flushes residuals out of the needle bore. This is illustrated in Fig. 2.23.

*On-column injection.* The injection modes previously described all require flash vaporization of the sample. As explained, this can lead to sample discrimination. At

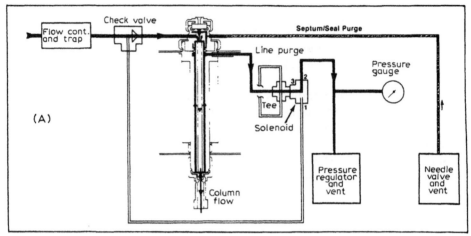

Splitless flow diagram, before injection

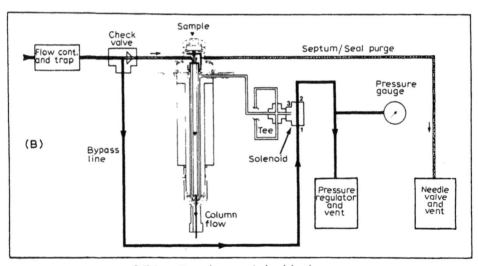

Splitless flow diagram, during injection

Fig. 2.22. Splitless mode of sample injection operation for WCOT columns. (A) Most carrier flow (dark line) acts to purge inlet. (B) During injection, activation of check valve and solenoid removes inlet purge while maintaining column flow. After injection, reactivation of inlet purge removes residual solvent to provide improved GC peak shape.

the injection temperature usually employed, it is also possible to observe decomposition of thermally labile compounds. These problems can be overcome by injecting the sample directly onto the WCOT column through a cool injection port (i.e., injection port is at same temperature as the column). Small volumes of solvent can be reproducibly injected since all of the sample reaches the column. As no sample splitting is performed, quantitative recovery of high-boiling sample components is

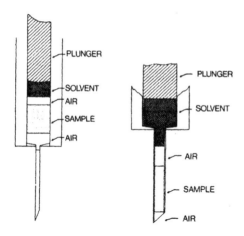

Fig. 2.23. Sandwich sampling technique.

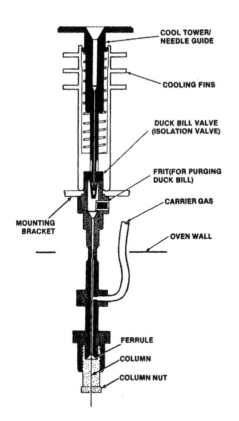

Fig. 2.24. On-column sample injector for WCOT columns.

32

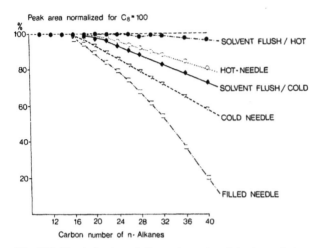

Fig. 2.25. Discrimination of alkanes by various injection techniques.

possible. This method of injection for WCOT columns was not prominent among early injector designs due to the mechanical difficulties of aligning the syringe needle with the column. Since typical inner diameters of WCOT columns are 0.25 mm (narrow bore) to 0.35 mm (wide bore), almost perfect needle alignment is required. Like splitless injection, on-column operation requires cold-trapping or the solvent effect to concentrate the sample at the head of the column.

Injector designs which are suitable for on-column work have only recently become available commercially. Fig. 2.24 shows one such design. In this injector, a duck-bill valve forms a gas-tight seal that forces carrier gas to flow through the column. When the plunger is depressed, gas flow is temporarily suspended and a metal guide is forced through the duck-bill valve. This allows the syringe needle to pass the valve and enter the column. When the plunger is released, gas flow through the column is restored. To obtain a very inert and perfectly straight syringe needle, a piece of fused silica is used instead of the usual metal needles. Although the on-column technique is gaining rapid acceptance, there are some limitations on its use. On-column automatic injectors are in the early stages of development, and will require different approaches than conventional autosamplers. Also, greater care must be taken in sample preparation, since non-volatile material in the sample will reach the column and accumulate at the column head. The comparative discrimination of various types of injection techniques is illustrated in Fig. 2.25.

### 2.5.2. Temperature programming

A very important analysis technique in gas chromatography is temperature programming (PTGC). This is used to improve separation and speed up analysis, especially for complex samples which contain many compounds having a wide range of boiling points. The technique is performed by changing the temperature of the GC column at a definite rate as the analysis proceeds. This causes components of

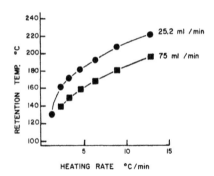

Fig. 2.26. Graphical illustration of relationship between heating rate, carrier gas flow rate, and retention temperature.

the sample to elute at an earlier time at a definite retention temperature than they would under isothermal conditions. The relation between the heating rate, carrier gas flow rate and retention temperature for a given compound is shown in Fig. 2.26. The advantage achieved in analytical data obtained is illustrated in Fig. 2.27. High molecular weight components that have high retention times and broad shapes in isothermal GC analysis elute as rapidly as the narrow shaped peaks using PTGC.

Fig. 2.28 illustrates the advantage of PTGC compared to isothermal analysis for a mixture of normal hydrocarbons. The effect of operating at a constant temperature which is optimum for early eluting peaks is that higher boiling components will take a very long time to elute, and their peak shape are poor. Some components may even remain on the column, or be eluted when the next sample is injected. One solution is to operate at a high initial temperature, as in Fig. 2.28 for the GC trace obtained at 150°C. Although high-boiling components are now eluted in a reasonable time and with good peak shape, low-boiling compounds have almost no retention on the liquid phase used and are not separated. The PTGC run shows all components of this mixture of a wide boiling range of compounds are eluted with good peak shape and in reasonable time. The GC is started at a low temperature that is effective for low-boiling compounds, then the oven temperature increased at a constant rate to a high final temperature.

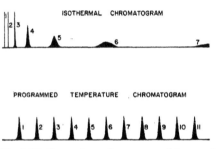

Fig. 2.27. Comparison of programmed temperature gas chromatography to isothermal.

34

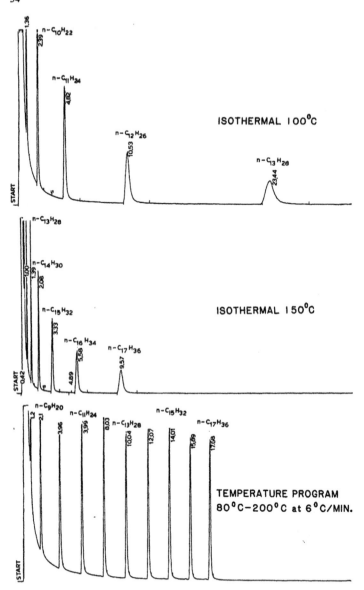

Fig. 2.28. Chromatograms of a mixture of hydrocarbons obtained by PTGC and isothermal (100 and 150 °C) GC.

One disadvantage with PTGC is that the stationary liquid phase on columns tends to bleed from the column as temperature is increased. This will result in an increasing background that may make reproducible and accurate peak areas difficult to obtain. For mass spectrometric detection, this column bleed will cause many extra or background ions to appear in the mass spectra of minor components.

Identifying the compound by its mass spectrum will be difficult. Although the problem of column bleed can be severe, proper choice of stationary phase, column conditioning before use, and using a reduced liquid film thickness will in most cases give low bleed. WCOT columns have been developed with immobilized stationary phases. By cross-linking and chemical bonding of the liquid film, very low column bleed is observed, even at elevated temperatures.

PTGC operation is necessary for effective operation of some of the injection techniques described previously. For splitless operation where the sample must be reconcentrated by cold trapping or the solvent effect, and for on-column operation in which the injector and initial column temperatures are both low, multi-ramp PTGC is generally employed. After injection, the GC oven is initially heated at a high rate to achieve rapid elution of the solvent, then this rate of heating is slowed down when the components of interest begin to elute. A typical PTGC analysis may be: initial temperature, 50°C; increase temperature at 15°C/min to to 150°C; change rate to 4°C/min until 250°C; hold at 250°C for 10 min. Except for instruments designed for specific applications not requiring PTGC, virtually all modern gas chromatographs are equipped to perform multi-ramp PTGC operation.

## 2.5.3. Retention index

If all experimental conditions could be held constant, then the retention times of compounds would also be constant for a specific column. A column whose characteristics did not change with time and that could separate all possible compounds could then be used for positive identification based solely on retention behaviour. Of course, such a column does not exist. By analyzing a compound on two or more columns having different stationary phases, however, the retention times can be used for positive identification when combined with information from other techniques. Unfortunately, small variations in conditions also affect retention times which makes peak identification difficult.

The solution to this problem is to measure retention times in relation to standards that are analyzed under the same conditions. These standards are assigned retention index (RI) values which by definition are constant. The RI value of a sample peak is determined by comparing its retention time to retention times of the standards which elute just before and just after the sample peak. Although small variations in conditions will cause sample retention times to shift, the retention times of standards will shift in the same manner and calculated RI values will be constant.

The first RI system developed and still the most widely used is the Kovats retention index ($RI_K$). In this system, normal alkanes are assigned an $RI_K$ value of $100 \times$ carbon number, where the carbon number is simply the number of carbon atoms in the hydrocarbon chain. Therefore, pentane ($C_5H_{12}$) will have an $RI_K = 500$, while hexadecane ($C_{16}H_{34}$) has an $RI_K = 1600$. For isothermal operation, $RI_K$ values are calculated according to the following expression:

$$RI_K = 100 \left[ \frac{\log t_i' - \log t_n'}{\log t_{n+1}' - \log t_n'} \right] + 100n$$

where $t_i'$ = corrected retention time of sample peak $(t_i - t_{air})$, $t_n'$ = corrected retention time of normal alkane with carbon number n, which elutes before the sample peak, and $t_{n+1}'$ = corrected retention time of normal alkane with carbon number $n + 1$, which elutes after the sample peak.

The values of $RI_K$ may be calculated using this equation or determined graphically as shown in Fig. 2.29. Retention times are corrected for dead volume by substracting the retention time of an unretained solute such as air or methane. Since this factor is constant for a specific system, and is a very small correction, it is often omitted except for the most precise determinations. It is necessary to specify the temperature and liquid phase used for reported $RI_K$ values, since only the normal alkanes have the same $RI_K$ values under all different conditions. An interlaboratory reproducibility of one $RI_K$ unit is feasible. Under linear temperature programming conditions (i.e. constant program rate) the above expression is simplified by omitting logarithms and using retention times directly.

One weakness of Kovats' system is that all types of compounds are not affected by changes in conditions to the same extent as normal alkanes. For this reason, RI systems have been developed using other homologous series of compounds for specific applications. For example, in the analysis of polar compounds a homologous series of alcohols such as pentanol, hexanol, and higher molecular weight alcohols, could be used.

A most effective way to obtain retention index values for polycyclic aromatic hydrocarbons is to use a series of aromatic hydrocarbons of increasing numbers of aromatic rings as reference compounds. By assigning a value of 200 to naphthalene, 300 to phenanthrene, 400 to chrysene and 500 to picene a linear and effective system can be set up for high molecular weight aromatic compounds.

Modern chromatographs are capable of achieving very reproducible conditions, even under PTGC operation. With a much improved technology for manufacture of reproducible WCOT columns, the value of the retention index for qualitative

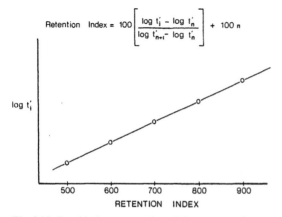

$$\text{Retention Index} = 100 \left[ \frac{\log t_i' - \log t_n'}{\log t_{n+i}' - \log t_n'} \right] + 100 \, n$$

Fig. 2.29. Graphical representation of Kovats retention index values.

analysis of complex samples has increased greatly, especially when RI data are used to supplement the information obtained from selective GC detectors.

## 2.6. Instrumentation

The principle of gas chromatography is so simple that a working chromatograph could be assembled in almost any laboratory using a few components. The first commercial chromatographs were rather crude models, but produced good data. Some of these are still in use today. As more powerful column technology, detectors and techniques such as PTGC were developed, the commercial instrumentation had to become more sophisticated in response to the needs of these developments. The gas chromatographs available today are marvels of precision and electronic sophistication. The greatest advances of all have been brought about by designing the chromatograph around a dedicated computer, instead of considering the computer as a data handling accessory.

### 2.6.1. Computerization

During the past few years, the cost of computers and associated equipment has dropped to the point where it is now feasible to dedicate a microprocessor to the operation and data analysis of a single gas chromatograph. Data processing advantages of computers are obvious, although a detailed understanding of the way data are handled and the flexibility inherent in this approach are necessary to maximize these benefits. Digital control of chromatographs and digital acquisition of data possess advantages which are less obvious since these functions are often transparent to the user. Before examining these benefits, it is important to understand the difference between analog and digital operation.

*Analog versus digital comparison.* The electrical output of a GC detector is a voltage that varies as a function of time. This voltage is proportional to some physical property being measured, such as the current produced by ionization of an eluting substance. Early analog instruments plotted this signal on a chart recorder after filtering the signal through electric circuits consisting of components such as resistors and capacitors. All data handling such as retention time and peak area determination was performed manually. Control of instrument operation in an analog instrument is performed by setting switches and knobs on a control panel. To check whether instrument conditions such as column temperature are properly set, a meter is generally provided. No permanent record of these settings is available unless recorded manually.

Instead of operating entirely with voltages controlled through electric circuits, these voltages are at some stage converted to representative numbers in a digital instrument. Electric circuits are still necessary to operate heaters, generate voltages from detectors, and operate other components, however, major functions such as instrument control, data acquisition, and data analysis are controlled through a computing device which makes decisions based on comparing numbers. Analog instruments can be interfaced to computers so that the detector signals are converted to numbers for data processing. Advantages of computerization are many for both instrument control and data acquisition and analysis.

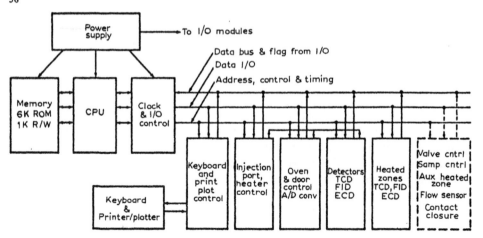

Fig. 2.30. Computerized GC system architecture: Hewlett-Packard 5830A instrument.

*Instrument control by microprocessor.* The Hewlett-Packard 5830A gas chromatograph was one of the first computerized chromatographs in which all GC operating conditions were entered through a keyboard. Fig. 2.30 illustrates the system architecture for this instrument and shows the many functions controlled by the microprocessor central processing unit (CPU). Directly associated with the CPU are the memory, clock, and data input/output (I/O) operations. Memory consists of permanent storage for pre-programmed functions and methods (ROM or read-only-memory) and temporary storage for chromatographic data such as the injection port temperature, initial GC temperature, program rate, and other functions (read/write memory). By adding additional memory and peripheral devices such as tape units or disk drives, storage for the entire chromatographic analysis can be obtained. The clock is important for internal timing functions which include the determination of accurate retention times.

Instrument control is provided by comparing measured parameters with the digital values entered by the operator and stored in memory. The measured values are voltages which are converted to numbers by an analog-to-digital (A/D) converter before being compared to the set value. If the measured value is not correct, action is taken by the CPU. Fig. 2.31 illustrates a typical operation, in which the GC oven is being set at the initial temperature (90°C) for analysis. If the temperature is correct, a ready light will be turned on. If not, action will be taken to change the temperature depending upon whether the temperature is high or low.

The actual programs that control instruments are much more complex than illustrated in Fig. 2.31. All of the heated zones, program rate and detector signals are monitored by the CPU. Accurate PTGC operation can be obtained because at hundreds of times during the analysis the correct oven temperature is calculated and compared to the measured value. The program rate can be changed at any time during the analysis by entering the new value through the keyboard. Set values can be listed in tabular format at the beginning or end of an analysis, while individual

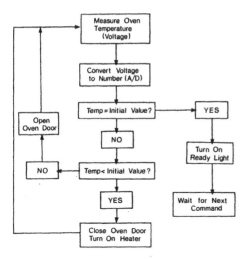

Fig. 2.31. Example of computerized control of GC in which the operations are controlled by YES/NO and ON/OFF commands.

values can be displayed when desired. Because all parameters are stored as numbers, peaks that are off-scale or chart recorder rate changes do not affect the accuracy of peak areas or retention times.

### 2.6.2. Data acquisition and analysis

An example of a chromatogram generated by a computerized instrument is shown in Fig. 2.32. Detector signals were transformed into numbers by an A/D converter and plotted by CPU control. The operating conditions were printed before beginning the analysis, and peak retention times are printed on the chromatogram as peaks are detected. Any changes made by the operator during the analysis such as attenuation and chart speed are also printed on the chromatogram.

At the end of the analysis, peak areas and retention times are listed. Some computerized instruments perform automatic quantitation of selected peaks by comparing peak areas with areas of standards injected before (external standardization) or at the same time (internal standardization) as the sample. Procedures can be quite complex and some instruments have the capability of user programming through a language such as BASIC to customize chromatographic reports. If the raw chromatographic data are stored in computer memory or on a peripheral storage device, peaks can be re-integrated using different parameters to achieve optimum results. Unattended operation can be performed by installing an autosampler, which is also controlled by the CPU.

Through computerization, greater quantities of more accurate chromatographic data can be obtained. Rapid method development, automatic report generation, and unattended automatic analysis allow operators to spend most of their time on other tasks. Sophistication of computerized instruments has reached the stage where the

40

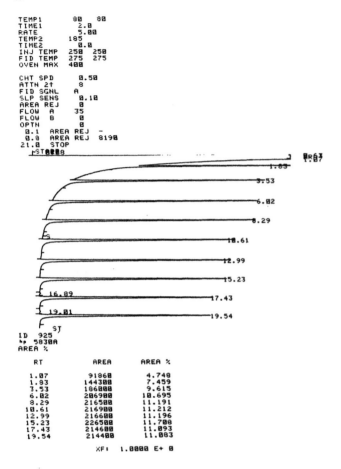

Fig. 2.32. Chromatogram generated by the plotter-printer of a computerized chromatograph.

major factors in many instrument purchases are the computer methods for displaying, analyzing and reporting data.

## 2.7. Suggested reading

1 F.I. Onuska and F.W. Karasek, *Open Tubular Column Gas Chromatography in Environmental Sciences*, Plenum Press, New York, 1984.

2 R.R. Freeman (Editor), *High Resolution Gas Chromatography*, Hewlett-Packard, 2nd ed., 1981.

3 K. Grob and G. Grob, *J. High Resolut. Chromatogr. Chromatogr. Commun.*, March (1979) 109–117.

4 W. Jennings, *Gas Chromatography with Glass Capillary Columns*, Academic Press, New York, 1978.

5 W.G. Jennings (Editor), *Applications of Glass Capillary Gas Chromatography (Chromatographic Science Series, Vol. 15)*, Marcel Dekker, New York, 1981.

6 H.M. McNair and E.J. Bonelli, *Basic Gas Chromatography*, Varian Aerograph, Walnut Creek, CA, 5th ed., 1969.

7 W.R. Supina, *The Packed Column in Gas Chromatography*, Supelco, Bellefonte, PA, 1974.

8 J. Sevcik, *Detectors in Gas Chromatography (Journal of Chromatography Library, Vol. 4)*, Elsevier, Amsterdam, 1976.

9 C.E. Reese, *J. Chromatogr. Sci.*, 18 (1980) 201–206.

CHAPTER 3

# MASS SPECTROMETRY

When a molecule is ionized in a vacuum, a characteristic group of ions of different masses are formed. When these ions are separated, the plot of their relative abundance versus mass constitutes a mass spectrum. This spectrum can be used to identify the molecule. Mass spectrometry first began with the work of physicists in 1908. By 1918 the technique had advanced to where it could be used to show that neon consisted of two isotopes of masses 20 and 22. This information revolutionized the concepts of chemical atomic masses and foretold the atomic age. By 1942 mass spectrometers were commercially available to perform analyses of mixtures of gases at remarkable speeds and accuracy. Today, mass spectrometry ranks as a major discipline in all aspects of scientific research.

## 3.1. Fundamentals

Mass spectrometry can be divided into two separate processes: ionization, and mass separation and recording of the ions formed. Different methods of ionization can be combined with the different techniques of mass separation depending upon the results desired.

### 3.1.1. Electron ionization

The schematic ionization chamber shown in Fig. 3.1 can be used to describe the principles and characteristics of ionization by electrons. The chamber will initially be at a very low pressure of $10^{-8}$ Torr. Electrons from a hot wire filament will be focussed across the chamber and attracted to an electrode having a potential of

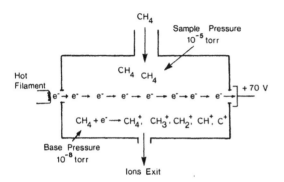

Fig. 3.1. Schematic of ionization chamber for producing ions by electron bombardment.

$$CH_4 + e^- \rightarrow \begin{matrix} CH_4^+ \\ CH_3^+ \\ CH_2^+ \\ CH^+ \\ C^+ \end{matrix}$$

Fig. 3.2. Ions produced by electron bombardment of methane.

70 V. This gives each electron an energy of 70 eV. When a sample such as $CH_4$ is introduced to the ionization chamber in sufficient quantity to increase the pressure to $10^{-5}$ Torr, collisions between the electrons and the $CH_4$ molecules causes a series of fragmentation reactions to occur. This results in production of all the positive ions conceivable from breaking bonds in the molecule, as illustrated in Fig. 3.2, because the 70 eV electrons have sufficient energy to break every bond in the molecule. Under the low pressure conditions involved, only about 1 molecule in $10^6$ undergoes collision and ionization. The $10^{-5}$ Torr sample pressure has an important consequence in producing representative and reproducible mass spectra. The mean free path (MFP) of an ion or molecule at that pressure exceeds the dimensions of the ion chamber, as given by the following approximate equation

$$MFP(cm) = \frac{5 \cdot 10^{-3}}{P}$$

where $P$ is the pressure (Torr). This means that the molecules and ions formed will behave independently as individuals and do not collide with each other, only with the chamber walls. This permits the withdrawal of ions, separation of ions of different masses, and ion collection at the detector without the disturbing effects of collision and reaction. The mass spectra thus produced are reproducible and characteristic of the molecule being ionized.

*Variables affecting ion formation.* The relative abundances of the ions formed by electron bombardment depend upon two primary variables: energy of the ionizing electrons and temperature at which ionization occurs. Fig. 3.3 illustrates the abundance of molecular and fragment ions as a function of the energy (voltage) of the ionizing electrons. As the energy increases from zero, a value will be reached around 10 V at which the molecular ion will just begin to appear. This is known as the appearance potential. As the energy of the electrons increases, the abundance of this molecular ion increases until its generation becomes relatively constant around 70 V. Between 10 and 15 V the fragment ions begin to appear and follow individual abundance curves as shown. Use of 70 V energy for ionizing electrons results in a mass spectrum created at a stable point in the formation of ions and gives mass spectra which are reproducible and characteristic of the molecule ionized.

Ion formation can be looked at as a reaction between an energetic electron and a molecule in which energy is transmitted to the molecule. This raises its energy to the

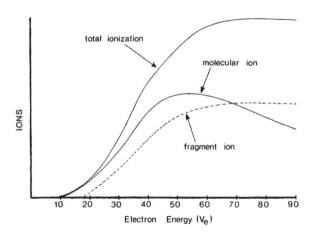

Fig. 3.3. Ion formation as a function of energy of ionizing electrons.

point where the molecule decomposes through a series of sequential reactions producing ions and neutral products. The extent of these reactions depends on the temperature at which they occur, therefore, the relative abundances of ions formed, which eventually are observed as the mass spectrum, is dependent on the temperature of formation. The degree of dependence is related to the molecular structure. For example, branched chain hydrocarbons display an increased dependence on temperature related to the degree of branching in the molecule. Comparison of unknown mass spectra with published mass spectra must take into account the temperature at which each was produced.

### 3.1.2. Chemical ionization

In electron impact (EI) mass spectrometry, the electron beam interacts with the sample molecules in the ion source to give a complex mixture of ions. The fragmentation ions produced can provide structural information and their relative abundances can be used for establishing compound identities. If the molecular weight was known in addition to this structural data, compound identification would be greatly facilitated. Under EI conditions, however, many molecules do not have stable molecular ions and the relative abundance of the molecular ion may be so small that positive identification cannot be made. This is especially true for homologous compounds such as straight chain alkanes, alkenes, or alcohols. It may be impossible to distinguish between long chain homologues based on the appearance of their respective EI mass spectra alone.

To complement the structural information obtained from EI mass spectrometry, a soft ionization technique can be employed to gain molecular weight information. Chemical ionization (CI) was introduced as an alternative ionization technique in 1966. Under EI conditions ions are formed by unimolecular processes whereas under CI conditions ion-molecule reactions are made to occur. In chemical ionization mass spectrometry (CIMS) a reagent gas is introduced into the ion source and a relatively high pressure is maintained (typically about 1 Torr). The reagent gas is ionized by the electron beam to produce reactant ions which can then interact with the sample molecules. Direct ionization of the sample molecules by the electron beam does not occur to any appreciable extent due to the relatively large concentration of reagent gas molecules. However, the probability of the sample interacting with the reactant ions is high because of the high pressure maintained in the source. Therefore, the ions appearing in the CI mass spectrum of a compound are due to ion-molecule reactions. Since these ion-molecule reactions are low in energy compared to the EI process, abundant molecular ions and simple fragmentation patterns are often observed.

CI mass spectra can be obtained on most commercially available spectrometers with only minor changes in the overall instrumentation. The ion source must be made as gas-tight as possible, and therefore, the electron beam aperture and the ion exit slit are usually reduced in area. Unfortunately, decreasing the area of these two slits results in a loss of sensitivity. In most commercial systems having EI/CI capabilities, it is possible to alter the ion source configuration without venting the system to the atmosphere. This is accomplished by the use of an ionization control selector which positions a tighter ion-source housing in place of that used for EI operation. With the exception of this modification, the mass spectrometer instrumentation is identical in EI and CI modes of operation.

A great deal of research has been carried out in which the formation of both positive and negative ions using a wide variety of reagent gases has been investigated. The majority of the work done has involved the study of positive ions. In recent years, increased interest in negative CI technique has been observed.

The most commonly used reagent gas in positive CI is methane which has been employed in the mass spectral analysis of many different organic compounds. The principal ions formed initially in the electron impact ionization of methane are $CH_4^+$, $CH_3^+$, and $CH_2^+$. These ions then react with other methane molecules to give the $CH_5^+$, $C_2H_5^+$, and $C_3H_5^+$ ions according to the reactions shown in Fig. 3.4. The reactions leading to the formation of reactant ions are all very rapid and the resulting ions are stable under CIMS conditions. The approximate relative abundances of the three major reactant ions $CH_4^+$, $C_2H_5^+$ and $C_3H_5^+$ are 48, 40 and 12%, respectively. Their relative abundances depend upon the source pressure and temperature.

The ions produced by the reactions discussed above act as Brønsted acids, $BH^+$, which may react with the sample molecules entering the ion source in several ways. The major reaction that occurs is proton transfer, which initially gives the quasi-molecular ion $[M + H]^+$.

$$BH^+ + M \rightarrow [M + H]^+ + B$$

This reaction is efficient if the proton affinity of the molecule, PA(M), is greater than the proton affinity of the reagent gas, PA(B). Under these conditions, the proton transfer reaction will be exothermic. The proton affinities of several Brønsted acid reagent gases are listed in Table 3.1. If PA(M) = 130.5 kcal mol$^{-1}$, then $CH_5^+$

PRIMARY STEP BY ELECTRON IONIZATION

$$CH_4 + e \longrightarrow CH_4^+, CH_3^+, CH_2^+, CH^+, C^+, H^+$$

SECONDARY STEP BY ION-MOLECULE REACTIONS

$$CH_4^+ + CH_4 \longrightarrow CH_3 + CH_5^+$$

$$CH_3^+ + CH_4 \longrightarrow H_2 + C_2H_5^+$$

$$C_2H_5^+ + CH_4 \longrightarrow 2H_2 + C_3H_5^+$$

Fig. 3.4. Formation of reactant ions in methane positive chemical ionization.

will react by exothermic proton transfer while PA(M) must be above 163.5 kcal $mol^{-1}$ for $C_2H_5^+$ to undergo proton transfer. Many organic molecules have proton affinities greater than 180 kcal $mol^{-1}$ and therefore most will be protonated in methane CI mass spectrometry. The extent of fragmentation increases with increasing exothermicity of the protonation reaction. In methane CI mass spectra, there are

TABLE 3.1
BRØNSTED ACID CI REAGENT GASES

| Reagent gas, B | Reactant ion, $BH^+$ | Proton affinity, PA(B) (kcal $mol^{-1}$) | Hydride ion affinity, HIA($BH^+$) (kcal $mol^{-1}$) |
|---|---|---|---|
| $H_2$ | $H_3^+$ | 100.7 | 300 |
| $N_2/H_2$ | $N_2H^+$ | 117.4 | 283 |
| $CO_2/H_2$ | $CO_2H^+$ | 128.6 | 272 |
| $N_2O/H_2$ | $N_2OH^+$ | 137.0 | 264 |
| $CO/H_2$ | $HCO^+$ | 141.4 | 259 |
| $CH_4$ | $CH_5^+$ | 130.5 | 270 |
| | $C_2H_5^+$ | 163.5 | 272 |
| $H_2O$ | $H^+(H_2O)_n^*$ | 173.0 | 227 |
| $CH_3OH$ | $H^+(CH_3OH)_n^*$ | 184.9 | 119 |
| $C_3H_8$ | $C_3H_7^+$ | 184.9 | 250 |
| iso-$C_4H_{10}$ | $C_4H_9^+$ | 196.9 | 231 |
| $NH_3$ | $N^+(NH_3)_n^*$ | 205.0 | 194 |

* Degree of solvation depends upon the partial pressure of the reagent gas (data given for $n = 1$).

46

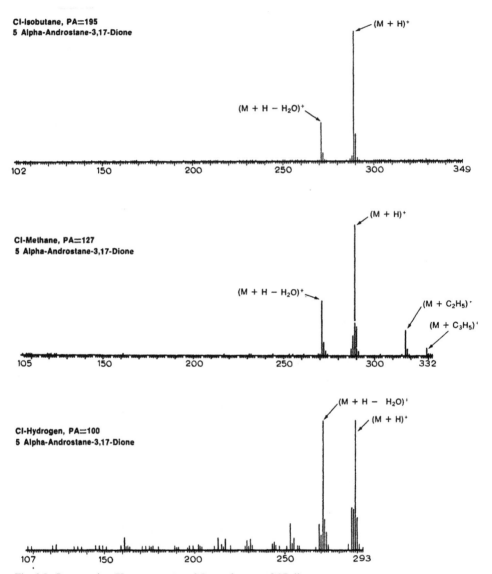

Fig. 3.5. Comparative CI mass spectra of 5-α-androstane-3,17-dione.

generally significant numbers of $[M + H]^+$ ions that remain unfragmented, while some structural information is still available from the limited fragmentation obtained. The major reactant ion formed in hydrogen CI, $H_3^+$, has a lower proton affinity than either $CH_5^+$ or $C_2H_5^+$ (see Table 3.1) and thus the protonation reaction is more exothermic for hydrogen CI than methane CI. Therefore, increased fragmentation is observed. The CI mass spectra of 5-α-androstane-3,17-dione using three different reagent gases are shown in Fig. 3.5.

The reactant ions formed in methane CI can also react by hydride abstraction as shown by the reaction give below.

$$BH^+ + M \rightarrow [M - H]^+ + BH_2 \; (or \; B + H_2)$$

Both the $CH_5^+$ and $C_2H_5^+$ ions have fairly high hydride ion affinities (HIA) as seen in Table 3.1. They therefore are capable of abstracting hydride ions from some organic molecules, giving $[M - H]^+$ ions in their respective CI mass spectra.

A third type of reaction is possible where charge exchange occurs between the reactant ion and the sample molecule.

$$BH^+ + M \rightarrow M^{+\cdot} + BH^{\cdot} \; (or \; B + H^{\cdot})$$

Ions resulting from the addition of $C_2H_5^+$ or $C_3H_5^+$ to the sample molecule are also sometimes observed in methane CI. These $[M + C_2H_5]^+$ and $[M + C_3H_5]^+$ ions can be very useful in determining the molecular weight of a compound. Both of these ions are generally much less intense than the quasimolecular ions observed, with the $C_2H_5^+$ cluster ion being the more abundant of the two.

In addition to the molecular weight information which can be obtained from the CI mass spectrum, it is also possible to gain some structural information from the fragmentation observed. For example, even-electron $[M + H]^+$ ions often fragment to form other even-electron ions by losing a stable molecule as shown in the reaction below.

$$RYH^+ \rightarrow R^+ + HY$$

These stable molecules, HY, may be $H_2O$, $CH_3OH$ (where Y = Cl, Br, I), or a variety of other small neutrals. Although the extent of fragmentation is generally much less than that observed in the corresponding EI mass spectra, these fragmentations are still useful in aiding with compound identifications.

Examples of EI and CI mass spectra of a variety of compound classes are given in Figs. 3.6–3.13. These spectra show the complementary nature of these techniques and help to illustrate the characteristics of CIMS.

Negative ions can also be studied under CI conditions. The instrumentation used is identical to that employed for positive CI studies, however the polarity of the various potentials must be reversed in order to look at negative ions. Negative ion reagent gas systems can be generalized into two classes, each of which is described below.

The simplest CI process which yields negative ions is electron capture. The capture of electrons by sample molecules is a resonance process which requires electrons of low energy. In order to obtain electrons with near-thermal energy, the CI reagent gas is used only as a moderating gas. The incident electrons, which may have energies as high as 240 eV for some instruments, are thermalized by inelastic

48

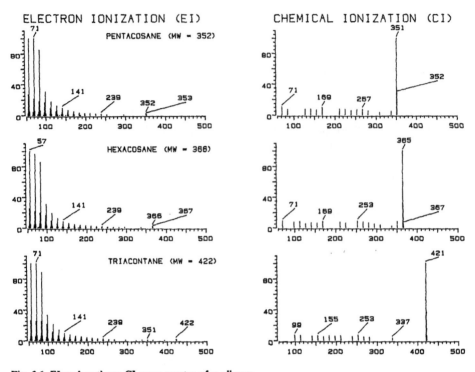

Fig. 3.6. EI and methane CI mass spectra of *n*-alkanes.

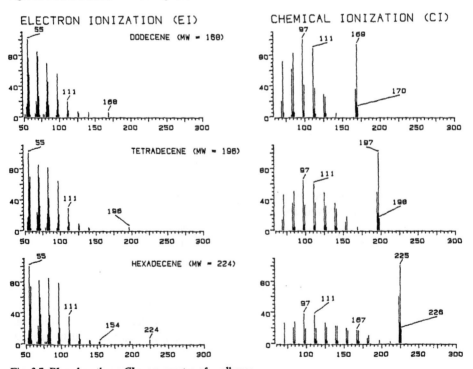

Fig. 3.7. EI and methane CI mass spectra of *n*-alkenes.

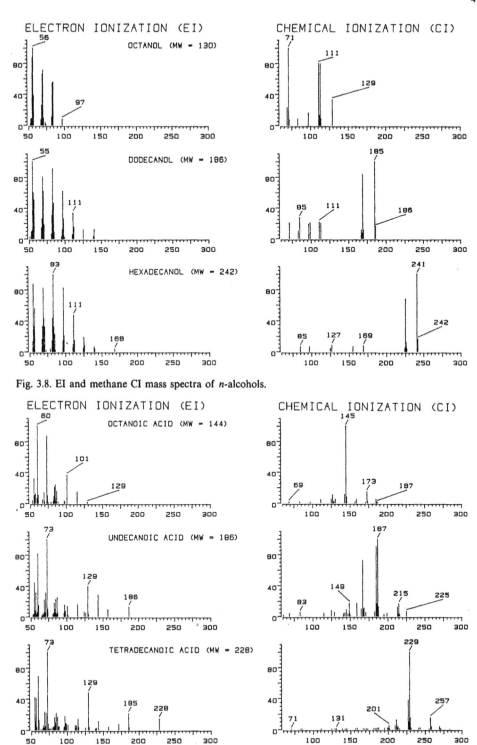

Fig. 3.8. EI and methane CI mass spectra of *n*-alcohols.

Fig. 3.9. EI and CI methane mass spectra of carboxylic acids.

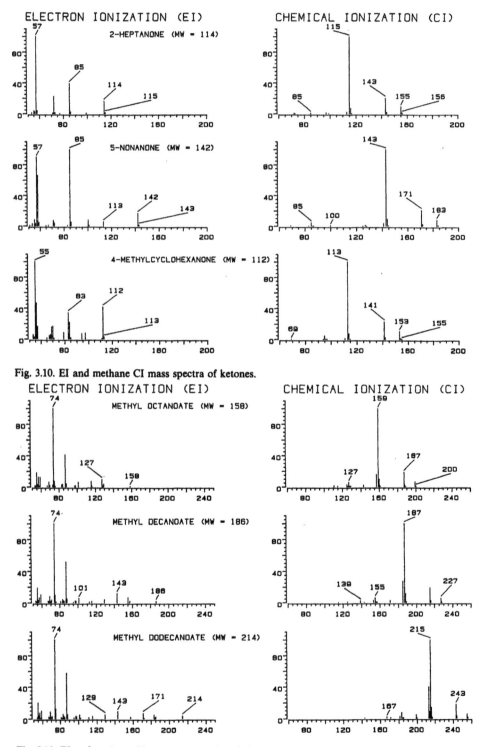

Fig. 3.10. EI and methane CI mass spectra of ketones.

Fig. 3.11. EI and methane CI mass spectra of methyl esters.

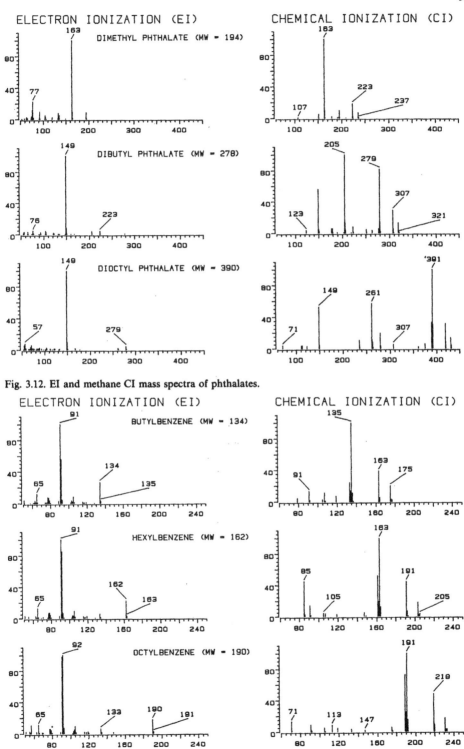

Fig. 3.12. EI and methane CI mass spectra of phthalates.

Fig. 3.13. EI and methane CI mass spectra of alkylbenzenes.

interactions with the reagent gas molecules. The electrons produced in the ionization of the reagent gas are also thermalized. Electron bombardment of the reagent gas therefore results in a high population of near-thermal energy electrons which can then be captured by sample molecules entering the ion source. Although nitrogen was usually used as the moderating gas in early experiments, reagent gases such as isobutane and methane are now commonly used.

The capture of an electron by a molecule can proceed by two different processes as shown below.

$$e^- + MX \rightarrow MX^- \qquad \text{non-dissociative electron capture}$$

$$e^- + MX \rightarrow M + X^- \qquad \text{dissociative electron capture}$$

The source temperature affects these two processes in opposite fashion. A decrease in the temperature will favor non-dissociative capture while an increase in the source temperature will enhance dissociative capture. The structure of the sample molecule is the major factor influencing which electron-capture mechanism is observed. While some molecules undergo predominantly non-dissociative electron capture, others may capture electrons by the dissociative process. Obviously the non-dissociative mechanism is much more desirable as it will yield molecular weight information whereas dissociative capture yields only information regarding the character of the anion $X^-$ which is not particularly useful.

Electron-capture negative CI is potentially useful for two reasons. The formation of molecular anions under CI conditions provides valuable molecular weight information for qualitative analyses. Negative CI often shows a significant increase in sensitivity for some molecules in comparison with both EI and positive CI. The rates of electron-capture reactions are generally much higher than those for ion-molecule reactions because of the greater mobility of the electron. However electron-capture CI is only applicable to molecules which have a positive electron affinity. Other molecules must be suitably derivatized in order to have a positive electron affinity. Therefore electron-capture CI is somewhat analogous to electron-capture detection as used in gas chromatography. It is also important that high-purity reagent gases be employed as the presence of oxygen-containing impurities result in irreproducible and often more complicated mass spectra.

Some examples of negative CI mass spectra using methane as the reactant gas are shown in Figs. 3.14–3.17 along with their corresponding EI and methane positive CI mass spectra.

Just as Brønsted acid reagent systems are used in positive CIMS, Brønsted base systems may be used for obtaining negative ions. The major reaction is proton abstraction as shown below.

$$B^- + M \rightarrow BH + [M - H]^-$$

This reaction is efficient if the proton affinity of the reactant anion, $PA(B^-)$, is greater than $PA([M - H]^-)$. In other words, BH must be a weaker acid than the

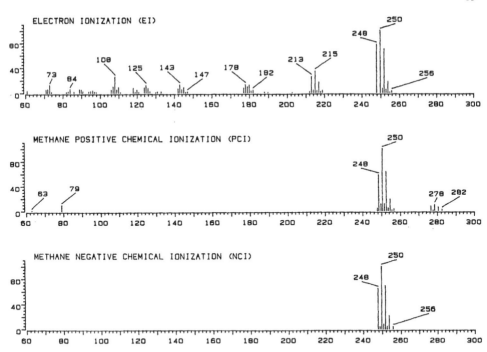

Fig. 3.14. EI, methane PCI and NCI mass spectra of pentachlorobenzene (MW = 248).

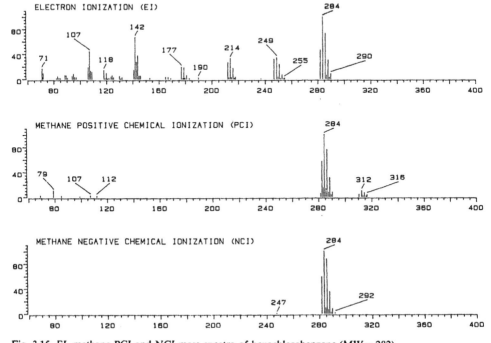

Fig. 3.15. EI, methane PCI and NCI mass spectra of hexachlorobenzene (MW = 282).

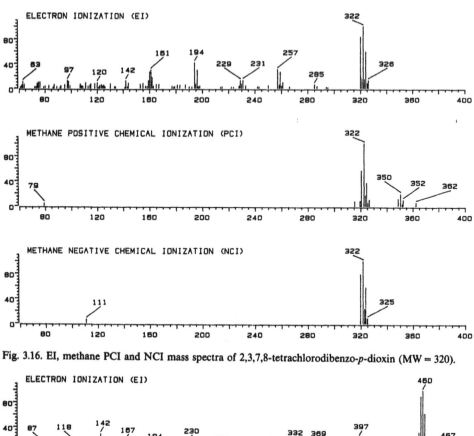

Fig. 3.16. EI, methane PCI and NCI mass spectra of 2,3,7,8-tetrachlorodibenzo-*p*-dioxin (MW = 320).

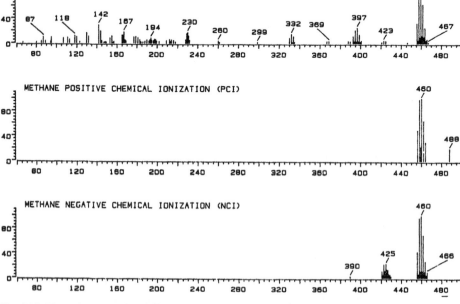

Fig. 3.17. EI, methane PCI and NCI mass spectra of octachlorodibenzo-*p*-dioxin (MW = 456).

TABLE 3.2
BRØNSTED BASE CI REAGENT SYSTEMS

| Reactant ion, B$^-$ | Proton affinity, PA(B$^-$) (kcal mol$^{-1}$) | Electron affinity, EA(B) (kcal mol$^{-1}$) |
|---|---|---|
| H$^-$ | 400 | 17.4 |
| NH$_2^-$ | 400 | 18.0 |
| OH$^-$ | 382 | 42.2 |
| O$^-$ | 382 | 33.7 |
| CH$_3$O$^-$ | 379 | 36.2 |
| F$^-$ | 372 | 78.4 |
| O$_2^-$ | 351 | 10.1 |
| Cl$^-$ | 333 | 83.4 |

sample molecule, M. Charge transfer reactions are also observed in cases where B has a sufficiently low electron affinity (EA). Species such as $CH_3O^-$ which have fairly high electron affinities generally do not undergo charge exchange reactions with most molecules. The proton and electron affinities for some common Brønsted base reagent gas systems are listed in Table 3.2. In most negative CI mass spectra using these types of reagents, little fragmentation of the $[M - H]^-$ ions is observed. This is believed to be due to the fact that most of the exothermicity of the reaction is associated with the B–H bond which is formed.

Although CI techniques in general are becoming more frequently used for a variety of mass spectral analyses, they have one common major disadvantage. The appearance of CI mass spectra depends very heavily on the ionizing conditions used. Factors such as the source temperature and pressure, the tuning of the various mass spectral parameters, and the cleanliness of the reagent gases and the ion source itself can drastically influence the results of CI. Therefore unlike EI, CI exhibits a considerable degree of irreproducibility. For example, it would not be possible to establish a mass spectral library of CI mass spectra for the purpose of compound identifications because the relative abundances of of the observed peaks depends so heavily on the exact ionization conditions. EI mass spectra are much more repro- ducible and can be used for compound identifications based on comparison of mass spectra. However, CI techniques when used with other techniques such as EI mass spectrometry or retention indices in gas chromatography–mass spectrometry (GC–MS) can be very valuable in establishing compound identities because of the molecular weight information which they provide. Proper choice of reagent gases can be used to increase sensitivity as is seen in electron-capture CI or to increase selectivity towards certain compounds or compound classes. Selected ion monitor- ing techniques employed in EI mass spectrometry can also be used for CI analyses and therefore considerable sensitivity, especially for negative ions, can be obtained. Although chemical ionization will not replace electron ionization, it is nonetheless a very valuable alternative ionization technique which has desirable characteristics for certain applications.

## 3.2. Instrumentation

Mass spectrometry is based on the fact that when a molecule is ionized in a vacuum, a characteristic group of ions of different masses are formed. A mass spectrum is produced by separating these ions and recording a plot of ion abundance versus ionic mass. A mass spectrometer consists of an ion source, a mass analyzer of ions, an ion detector, and a vacuum system.

Mass spectrometers are classified according to the principle used to separate ionic masses. The most commonly used mass spectrometers fall into two broad classes: quadrupole and magnetic. Either mass spectrometer can be interfaced to a gas chromatograph.

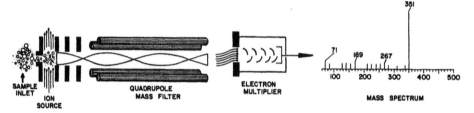

Fig. 3.18. Design of the quadrupole mass spectrometer.

### 3.2.1. Quadrupole mass spectrometers

The quadrupole mass spectrometer design is illustrated in Fig. 3.18. In the quadrupole mass spectrometer ion separation is accomplished by passing the ion beam through the centre of four parallel rods to which voltage is applied. The theoretically correct cross-sectional shape of these rods as a hyperbola. This provides the correct electrical field in the centre of the four rods. The principle of mass selection by varying this electric field is illustrated in Fig. 3.19. It is difficult to manufacture rods with hyperbolic surfaces of the necessary tolerances ($\pm 0.00001$

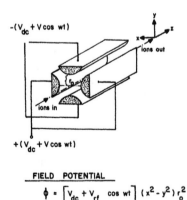

FIELD POTENTIAL

$$\phi = \left[ V_{dc} + V_{rf} \; \cos \; wt \right] (x^2 - y^2) \, r_o^2$$

Fig. 3.19. Principle of operation for quadrupole mass spectrometers.

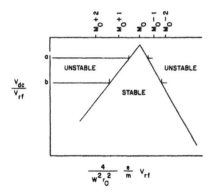

Fig. 3.20. Stability diagram for quadrupole mass spectrometry.

in.). Some quadrupole mass spectrometers use round rods to simulate a hyperbolic electrical field. In quadrupole instruments, there is a loss of sensitivity at high masses. Some of this is due to the effect of the fringing fields between the ion source and the rod structure; some is due to the use of round rods.

The separation of the masses in a quadrupole mass analyzer is accomplished by simultaneously applying both a dc and a radio frequency (RF) ac voltage to the rods. Mass resolution is determined by the ratio of the dc to RF voltage. Resolution is normally adjusted so that an ion of unit mass will pass completely through the rods to the detector. The stability diagram seen in Fig. 3.20 indicates the parameters involved in resolution. To obtain a mass scan the dc and RF voltages are varied while maintaining a constant dc:RF ratio. The mass permitted to pass through is linearly related to the amplitude of the voltage. This simplifies GC–MS operation as well as computerization of the system.

The linear relationship between mass and voltage makes control and calibration by computers easy. The ion trajectory through the rods follows an extremely complex oscillatory path which can be expressed mathematically. Quadrupole mass spectrometers have a reputation for high sensitivity and the ability to scan rapidly at millisecond intervals. These qualities are well suited for coupling with a gas chromatograph, especially when WCOT columns with their narrow peaks are used.

### 3.2.2. Magnetic sector mass spectrometers

The magnetic sector mass spectrometer is one of the two types most commonly used in GC–MS systems. Magnetic sector mass spectrometry dates back to the late 1800s and is the type of mass spectrometer on which all early work was performed. A magnet is used to separate ions for subsequent mass detection. Ions with a unit charge are influenced by the magnetic field according to the equation

$$m/z = H^2R^2/2V$$

where $m$ is the mass of the ion, $z$ is its charge, $H$ is the strength of the magnetic

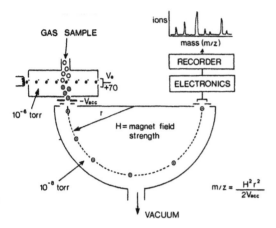

Fig. 3.21. Design of a magnetic mass spectrometer.

field in gauss, $R$ is the radius of curvature, and $V$ is the accelerating voltage applied to the ions. A mass scan can be done by varying either $H$ or $V$ in time. For most applications, $H$ is varied while maintaining a constant $V$. This type of instrument is schematically shown in Fig. 3.21.

An ion is accelerated from the ion source with a great deal of energy. This is accomplished through a repeller plate at 2000–8000 V. The magnetic field of the electromagnet is then scanned. The mass which passes through the analyzer and is focussed at the detector is dependent upon the square of the magnetic field strength. The ions leaving the source of the magnetic sector instrument all have a common kinetic energy determined according to the equation:

$$Ve = \tfrac{1}{2}mv^2$$

where $V$ is the accelerating voltage, $m$ is the mass and $v$ is their velocity. However, the ions have slightly differing initial velocities due to kinetics of ionization. The magnetic field then focusses the ions into a narrow beam, and they pass through a slit just prior to the detector.

If only a magnet is used, the instrument is called a single focussing instrument. Electrostatic field electrodes can be applied prior to the magnet, but after the ion source, to provide an initial focus to the ions exiting the source and remove the effect of their having different initial velocities. The instrument is then called a double focussing mass spectrometer. The ion beam must be double focussed in such a manner to perform high resolution mass spectrometry.

Fig. 3.22 shows how the resolution ($R$) of two mass peaks barely resolved is calculated. In this standard method of calculating $R$, barely resolved means that there is a slight overlap in the peaks causing the valley between peaks to be 10% above baseline. Exact molecular weights can be determined with high resolution mass spectrometers by measuring the degree of separation of sample ion peaks from ion peaks of a standard compound led continuously into the ion source whose exact

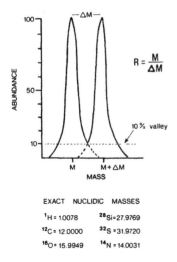

Fig. 3.22. Definition of resolution in mass spectrometry.

masses are accurately known. Quadrupole mass spectrometers operate by maintaining constant $\Delta m$ for its working range, which means that resolution is variable. Magnetic sector instruments, however, have constant resolution and variable $\Delta m$. This is illustrated schematically by Fig. 3.23.

The concept of exact molecular weight is important to mass spectrometry, and results from the fact that exact nuclidic masses are not simple integers. This is shown in Fig. 3.22, where all of the exact masses given are based on a universal standard taken to be $^{12}C = 12.000$. True high resolution GC–MS ($R \geq 20,000$) is

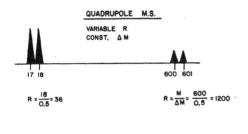

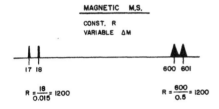

Fig. 3.23. Relationship between $\Delta m$ and resolution for quadrupole and magnetic mass spectrometers.

seldom performed for GC–MS applications, especially for WCOT column GC–MS, because accurate mass determination requires slow scan speeds and a stable high concentration of organic molecules in the ion source. Most GC–MS work using magnetic sector instruments is in the medium resolution (5000–10,000) range. Quadrupole instruments are used when only nominal or integer masses are required. This is often erroneously called unit resolution.

One of the advantages of magnetic mass spectrometers for GC–MS is that most established reference spectra were obtained with magnetic mass spectrometers. In addition, magnetic spectrometers scan a broader mass range (2–2000). Their scan rate is slower and limited because of magnetic field hysteresis. Recent development of laminated core magnets now permit scan rates almost as fast as quadrupole instruments.

### 3.3. Interpretation of mass spectra

A mass spectrum contains structural information of the molecule. The spectrum is plotted in a normalized bar graph form in which the abundances of all ions are displayed as percentages of the most abundant ion set to a value of 100%. A detailed interpretation of the mass spectrum can be accomplished following some well established fragmentation mechanisms and rules. In some cases the interpretation will result in a positive identification of the molecule. In others it will only give a partial solution to the identification problem and alternate methods must be used. Other methods commonly used include use of high resolution mass spectrometry for elemental determination, GC retention behaviour, infrared spectroscopy and NMR spectroscopy.

### 3.3.1. Types of ions

Electron ionization produces several different types of ions. These are indicated in Table 3.3. The identification of the molecular ion is the most important step in the interpretation of the mass spectrum. This will provide the molecular weight of the molecule. The relative abundance of the molecular ion is related to its stability,

TABLE 3.3
TYPES OF POSITIVE IONS FORMED IN ELECTRON IONIZATION

| Ion | Description |
| --- | --- |
| Molecular | The molecule with a positive charge by loss of electron |
| Base | The most abundant ion in the spectrum |
| Fragment | Formed by rupture of one or more bonds in the molecule |
| Rearrangement | Formed by rupture of bonds and migration of atoms |
| Doubly charged | Ions containing two positive charges by loss of two electrons appearing at one-half mass |
| Metastable | Ion ($m_1$) which fragments into ion of lower mass ($m_2$) and neutral particle will be observed at an ionic mass of m*. $m^* = (m_2)^2/m_1$ during transit of mass analyzer. Not observed in quadrupole systems. |

McLAFFERTY REARRANGEMENT

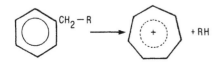

FORMATION OF TROPYLIUM ION

Fig. 3.24. Common rearrangement ions.

which depends upon its structure. For example, normal hydrocarbons have molecular ions of decreasing abundance with increasing molecular weight, while aromatic hydrocarbons have stable abundant molecular ions. This is because aromatic hydrocarbons are better able to stabilize a positive charge than aliphatic hydrocarbons.

The base ion represents the most stable ion in the spectrum and its mass can be used in a very diagnostic manner. It can be a molecular or fragment ion. There are tables of prominent ions and the possible structures related to them that can be used for guidance (see ref. 1, section 3.4).

Fragment ions give a great deal of information about the molecule since they are essentially the pieces of a molecule that can be arranged in some logical fashion to re-create the original structure.

There are characteristic rearrangement ions that indicate a definite structural entity exists in the original molecule. Two of the best known are the tropylium ion and McLafferty rearrangement shown in Fig. 3.24.

Doubly charged ions are diagnostic in that they commonly occur in certain types of compounds. High molecular weight and highly aromatic compounds display numerous doubly charged ions.

Metastable ions provide much structural information by giving the information on the decomposition pathways by which they are formed. They give diffuse ion peaks and are not observed in quadrupole mass spectrometers.

### 3.3.2. Isotopic abundances and characteristic ion clusters

Many of the elements exist in nature as several naturally occurring isotopes. The relative abundances of these isotopes are in fixed ratios, therefore molecules are composed of the isotopes of its component elements in the same ratios. An individual molecule will have a fixed molecular weight, but a large number of these same molecules will contain a distribution of molecular weights, depending upon which isotopes of individual atoms are present. A mass spectrum shows the presence of isotopes by giving more than one peak for each type of ion formed. The number

TABLE 3.4
RELATIVE ISOTOPIC ABUNDANCES OF ELEMENTS

| Isotope | Relative abundance |
|---------|--------------------|
| $^{12}C$ | 100 |
| $^{13}C$ | 1.1 |
| $^{14}N$ | 100 |
| $^{15}N$ | 0.40 |
| $^{16}O$ | 100 |
| $^{18}O$ | 0.20 |
| $^{19}F$ | Monoisotopic |
| $^{35}Cl$ | 100 |
| $^{37}Cl$ | 32.5 |
| $^{79}Br$ | 100 |
| $^{81}Br$ | 98 |
| $^{127}I$ | Monoisotopic |
| $^{31}P$ | Monoisotopic |
| $^{32}S$ | 100 |
| $^{34}S$ | 4.4 |
| $^{28}Si$ | 100 |
| $^{29}Si$ | 5.1 |
| $^{30}Si$ | 3.4 |

of extra peaks detected for each ion, and their relative ratios, depend only on the number and type of atoms in the ion and the relative ratios in nature of isotopes of these atoms.

Table 3.4 lists the relative abundances of isotopes of the more commonly found elements. These natural relative abundances are very useful in the interpretation of a mass spectrum. For example, the number of carbon atoms in a hydrocarbon molecule can be postulated by the abundance of the ion one mass unit higher than the molecular ion. This ion has the same structure but contains $^{13}C$ atoms. Its relative abundance will be 1.1 times the number of carbon atoms.

Some of the most important diagnostic ion clusters are due to the presence of chlorine and bromine atoms. These atoms produce easily observable cluster patterns because of the large abundances of their isotopes. In a random sampling of molecules of a compound containing C, H and one chlorine atom, for example, all of the chlorine containing ions will produce two peaks in the mass spectrum. One of the peaks will be two mass units higher, due to the $^{37}Cl$ isotope. Because its natural abundance is about 25%, while that of the $^{35}Cl$ isotope is 75%, the higher mass ion will have an abundance 33% of the $^{35}Cl$-containing ion. If the molecule contained two chlorine atoms, the higher mass ion relative abundance would be twice as great, because the probability of finding the $^{37}Cl$-isotope in an individual molecule would be doubled. Elements such as chlorine and bromine that have abundant stable isotopes two mass units apart are referred to as "A + 2" elements, while elements with predominant stable isotopes one mass unit apart are called "A" elements.

The abundances of isotope peaks can be calculated by a relatively simple formula. For a single atom containing two principal isotopes the expression is

$$(a+b)^n$$

where $a$ represents the light isotope, $b$ is the heavier, and $n$ is the number of atoms of the element in the molecule. Consider chlorine which has two principal isotopes, $^{35}$Cl and $^{37}$Cl, with relative fractional abundances of $\frac{3}{4}$ and $\frac{1}{4}$, respectively. If an ion containing two chlorine atoms has mass $M$ ($^{35}$Cl + $^{35}$Cl) then there will also be a peak in the mass spectrum at $M+2$ ($^{35}$Cl + $^{37}$Cl) and at M + 4 ($^{37}$Cl + $^{37}$Cl). The above expression for $n = 2$ gives

$$a^2 + 2ab + b^2$$

The three terms represent the fractional abundances of the $M$, $M+2$, and $M+4$ ions, respectively. They are obtained by substituting the natural fractional abundances for $a$ and $b$, which gives ratios of 9:6:1 for the $M:M+2:M+4$ peaks.

The same procedure can be used when more than one element with an abundant natural isotope is present. For an ion containing two elements, each of which has two naturally occurring isotopes of fractional relative abundances $a,b$ (element A) and $c,d$ (element B), the relative abundances of isotope peaks in the mass spectrum will be given by

$$(a+b)^2(c+d)^2$$

The above calculation becomes very tedius as more atoms are added, but it is easily performed by a computer. The simplest way to check the identity of a given set of clusters in a mass spectrum is to use graphical information such as shown in Fig. 3.25.

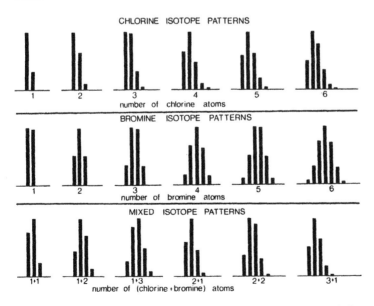

Fig. 3.25. Relative isotopic abundances produced by the presence of chlorine and bromine atoms in an ion.

### 3.3.3. Nitrogen rule and rings-plus-double-bonds

Because of the fixed valences of common elements, a few simple rules can be used to aid in the interpretation of mass spectra. The nitrogen rule is very useful, and is applicable to all compounds that have only covalent bonds and are composed of C, H, O, N, S, P, Si, As, the halogens, and the alkaline earth metals. In this list of elements, only nitrogen has an even mass and an odd valence. Therefore, all organic compounds composed of these elements must have an even molecular weight if an even number of nitrogen atoms (including zero) are present. Compounds that have an odd number of nitrogens must have an odd molecular weight. A further extension of this rule is that cleavage of a single bond gives an odd mass ion fragment from an even molecular ion, and an even mass ion fragment from an odd mass molecular ion, if all the nitrogen atoms present in the molecular ion are retained in the fragment ion.

Another consequence of the fixed valences of common elements is that the number of rings plus double bonds ($R$) of molecules can be calculated using the following formula:

$$R = x - (y/2) + (z/2) + 1$$

where $x$, $y$, $z$ are the numbers of carbon, hydrogen and nitrogen atoms in the molecular formula, respectively. In some cases the calculated value is non-integer. When this occurs $R$ is rounded off to the next lowest integer. Of course, $R$ is only useful if the molecular formula is known.

### 3.3.4. Steps in interpretation

Although there is no firm method established by which a skilled individual will examine and interpret a mass spectrum, there are a number of steps that can logically be used. These steps are listed in Table 3.5 and can be followed as far as necessary to complete the interpretation.

If the molecular ion can be identified much interpretative information is then available. Use of isotopic abundances can give some clues to the numbers and types of atoms making up the empirical formula of the molecule.

The general appearance of the spectrum can give structural clues. There are the

TABLE 3.5
STEPS IN THE INTERPRETATION OF A MASS SPECTRUM

1. Look for molecular ions
2. Note general appearance of spectrum
3. Scan spectrum for peak clusters of characteristic isotopic patterns
4. Look for small-mass neutral fragments lost from molecular ion
5. Look for characteristic low-mass fragment ions
6. Compare spectrum to reference spectra
   (a) using compilations in atlas
   (b) using computerized search of reference spectra
7. Interpret spectra using all information possible
8. Check intepretation by obtaining spectrum of pure compound under same instrumental conditions

characteristic patterns for normal alkanes in which there is a high abundance of fragment ions which decrease steadily with molecular weight. On the other hand aromatic compounds show high abundances of molecular ions and lesser amounts of fragment ions.

The loss of small mass neutral fragments from the molecule can indicate the presence of certain groups; a loss of 15 a.m.u. indicating a methyl group $(CH_3)$ and the loss of 18 a.m.u. indicating water $(H_2O)$. Tables listing the neutral fragments possibly associated with specific mass losses are useful in diagnosis.

There are many characteristic low mass fragment ions which can be observed by examining the spectra of numerous compounds. In many cases these ions form a well recognized series, such as the 43, 57, 71, 85 ions for the normal alkanes.

The comparison of a spectrum to compilations of reference spectra of known compounds can result in a rapid identification of the compound if its reference spectrum is available. These identifications can be ambiguous if isomers are involved. Computerized search systems range from a simple spectral matching to those providing interpretative information about the molecule.

If the interpretation of the mass spectrum is not completed by the end of step 6 of Table 3.5 then other information must be brought to bear on the problem. If there is sufficient sample, NMR and infrared spectroscopy can add valuable data by providing additional structural information. After a tentative identification has been made, positive GC–MS identification can only be assured by obtaining the mass spectrum and retention time of the pure compound under identical conditions.

### 3.3.5. Examples

The following examples illustrate the application of the concepts and procedures discussed above. Some simple molecules have been chosen to serve as examples of a few of the classes of compounds that may be encountered: alkanes, alkenes, halogenated hydrocarbons, alcohols, carboxylic acids and aromatic compounds. It is not intended that these examples and the practice problems which follow provide a comprehensive training in mass spectrum interpretation. Rather, they serve to show how the orderly application of the interpretation procedure can lead to the identification of these compounds. Interpretation can be facilitated by use of ref. 2, section 3.4.

*Example 1 (Fig. 3.26).* The peak at $m/z$ 32 is a likely candidate for the molecular ion. Peaks at $m/z$ 31, 29 and 15 give logical neutral losses $([M - 1]^+, [M - 3]^+, [M - 17]^+)$ for that assignment. The abundance of $m/z$ 33 relative to $m/z$ 32 is 1.5%, indicative of one carbon atom. There is no $[M + 2]^+$ peak, but oxygen may be present. For a molar mass of 32, $O_2$, or $CH_4O$ are possibilities. We can discount $O_2$ from the complexity and appearance of the spectrum. $CH_3OH$ (methanol) is the only possible structure. The $[M - 1]^+$ and $[M - 3]^+$ peaks represent losses of H and $H_3$, respectively. The $[M - 17]^+$ peak represents the loss of OH.

*Example 2 (Fig. 3.27).* The peak at $m/z$ 28 is the best choice for the molecular ion. The relative abundance of $m/z$ 29 is 2.3% compared to $m/z$ 28, which indicates two carbon atoms. $C_2H_4$ is the only logical formula. The $R$ value is 1, therefore $H_2C = CH_2$ (ethene) is the proper identification. The peaks at $m/z$ 24 to $m/z$ 27 represent the loss of hydrogen atoms.

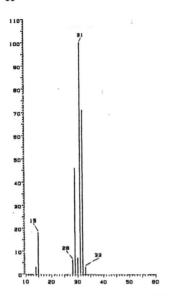

Fig. 3.26. Example 1: methanol ($CH_3OH$).

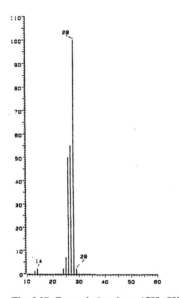

Fig. 3.27. Example 2: ethene ($CH_2CH_2$).

*Example 3 (Fig. 3.28)*. The odd mass of the molecular ion *m/z* 27 indicates an odd number of nitrogen atoms, according to the nitrogen rule. The low mass precludes more than one nitrogen atom, so the rest of the mass should be made of hydrogen and carbon. CHN is a likely formula. *R* is equal to 2, so the only molecular structure possible is HC≡N (hydrogen cyanide).

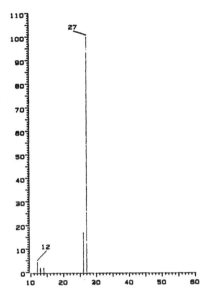

Fig. 3.28. Example 3: hydrogen cyanide (HCN).

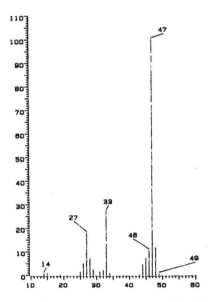

Fig. 3.29. Example 4: fluoroethane ($CH_3CH_2F$).

*Example 4 (Fig. 3.29).* On first inspection the $m/z$ 47 peak might be considered to be the molecular ion. However, the next major peak at $m/z$ 33 would represent a neutral loss of 14 a.m.u. from the molecular ion, which is highly unlikely. Also, the abundant even electron fragments $m/z$ 33 and 27 have odd masses, which suggest the molecular ion has an even mass. The peak at $m/z$ 48 is the molecular ion. The

68

relative abundance of $m/z$ 49 is 2.6% compared to $m/z$ 48, indicating a maximum of two carbon atoms. The spectral pattern indicates no A + 2 elements are present, so the remainder of the mass is made up of A elements or oxygen. Phosphorus and iodine are too heavy so the only possible formulas are $CH_4O_2$, $C_2H_5F$ and CHOF. CHOF is eliminated as a possibility because a neutral loss of 15 a.m.u. from the molecular ion is impossible from the structure. The ratio $(m/z\ 34)/(m/z\ 33)$ is 1.9%, indicating one carbon atom. The best assignment is $CH_2F^+$. The large $[M-1]^+$ peak at $m/z$ 47 is characteristic of aliphatic fluorines and has the general structure $R_2C{=}F^+$. The molecule is $CH_3CH_2F$ (fluoroethane).

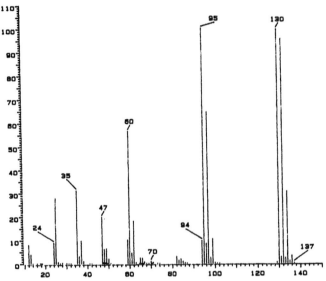

Fig. 3.30. Example 5: trichloroethene ($Cl_2C = CHCl$).

*Example 5 (Fig. 3.30).* This spectrum is an example of how contributions from A + 2 elements in an $[M-1]^+$ peak can affect isotope abundances of A + 1 elements in $[M]^+$ peaks.

The cluster of peaks at the high end of the spectrum suggest the presence of A + 2 elements, Cl or Br, in the molecule. The ratio of the peak intensities $(m/z\ 132)/(m/z\ 130)$ is 0.97 which could indicate one bromine atom but the mass peaks above $m/z$ 132 rule this out. The series of high intensity peaks at $m/z$ 130, 132, 134 and 136 is probably the $[M]^+$ isotope cluster, while the series of low abundances at $m/z$ 129, 131, 133 and 135 is likely the $[M-1]^+$ cluster. We should note that the $^{13}C$ peaks in each cluster will interfere with each other.

The group ratios are close to what would be expected for $Cl_3$. The expected abundances are calculated from $(1+0.325)^3$ and normalized to 100: A = 100, A + 2 = 98, A + 4 = 32, A + 6 = 3. The number of carbon atoms is best calculated using the highest peak in the Cl cluster.The ratio $(m/z\ 137)/(m/z\ 136)$ is 3.2% $(^{37}Cl_3C_n)$ which gives a maximum number of three carbon atoms. But more information is available. We have assumed the $m/z$ 136 peak is the A + 6 peak for the $^{37}Cl_3$ cluster which leads to the assignment of $m/z$ 130 as the molecular ion. Three chlorine atoms would account for 105 a.m.u. so 25 a.m.u. are left to be made up with A and A + 1 elements. $NH_{11}$ and $C_2H$ are possible, the latter most probable. A possible formula is $C_2HCl_3$. The $R$ value is 1, so the compound is likely $Cl_2C = CHCl$ (trichloroethene).

To confirm, the peak intensities of the isotope clusters can be calculated for $C_2Cl_3$ and $C_2HCl_3$ with the formula $(1 + 0.011)^2(1 + 0.325)^3$.The other clusters around $m/z$ 95 and 60 would be $[M - Cl]^+$ and $[M - Cl_2]^+$ respectively.

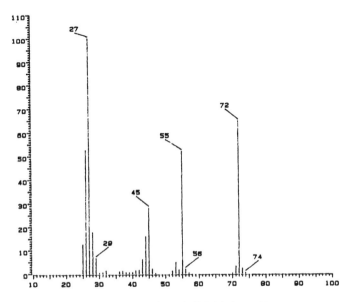

Fig. 3.31. Example 6: 2-propenoic acid (HOOC-CH = $CH_2$).

*Example 6 (Fig. 3.31).* The peak at $m/z$ 72 is a good candidate for the molecular ion. The major neutral losses represented by $m/z$ 55 ($[M - 17]^+$), $m/z$ 45 ($[M - 27]^+$) and $m/z$ 27 ($[M - 45]^+$) make sense. The small ratio $(m/z\ 74)/(m/z\ 72)$ indicates that no A + 2 elements except oxygen are present. The ratio $(m/z\ 73)/(m/z\ 72)$ is 4.2%. Allowing for error, this could represent three or four carbon atoms. The ratio $(m/z\ 73)/(m/z\ 72)$ is 0.7% which suggests the presence of oxygen atoms, since carbon atoms would account for only 0.05% of the abundance at A + 2. Therefore the possible molecular formulas are $C_3H_4O_2$ and $C_4H_8O$. Peak ratios can be calculated for these two formulas:

| $m/z$ | 72 | 73 | 74 |
|---|---|---|---|
| Observed | 100 | 4.2 | 0.7 |
| $C_3H_4O_2$ | 100 | 3.3 | 0.5 |
| $C_4H_8O$ | 100 | 4.4 | 0.3 |

The two proposed formulas are still possible, based on these ratios. The $R$ value for $C_3H_4O_2$ is 2, and for $C_4H_8O$ is 1. Possible structures are $CH_2$=CHCOOH and CHO-$C_3H_7$.

We have to look at the rest of the spectrum to distinguish between the two structures. An aldehyde spectrum should contain a large $m/z$ 29 (HCO) peak, which in this spectrum is only 5% abundance. The major peak at $m/z$ 55 represents a loss of OH from the molecular ion. Since an aldehyde cannot lose OH easily, we are left with $C_3H_4O_2$ as the molecular formula and $CH_2$ = CHCOOH (2-propenoic acid) as the identification.

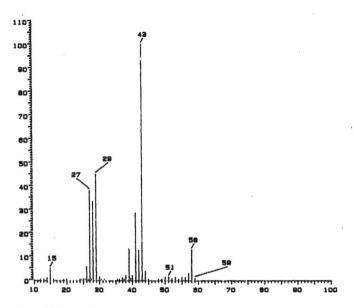

Fig. 3.32. Example 7: n-butane ($CH_3CH_2CH_2CH_3$).

*Example 7 (Fig. 3.32).* The molecular ion is $m/z$ 58. The ratio $(m/z$ 59)/($m/z$ 58) is 4%. This could indicate a maximum of four carbon atoms, but three carbon atoms is a possibility. The $m/z$ 43 base peak could have three carbon atoms, based on the $(m/z$ 44)/($m/z$ 43) ratio of 3.4%. This rules out an assignment of $C_2H_3O^+$ for the $m/z$ 43 peak. This peak can be assigned as $C_3H_7^+$, leaving the assignment of $C_4H_{10}$ to $m/z$ 58. To distinguish between n-butane, $CH_3CH_2CH_2CH_3$ and 2-methylpropane, $(CH_3)_3CH$, we must look at the lower mass fragments. 2-Methylpropane looses a methyl group to form $(CH_3)_2CH^+$. The ion $(CH_3)_2C^+$ ($m/z$ 42) is very stable, resulting from a loss of the hydrogen atom at the branch site. This peak is weak, but we note that the $m/z$ 29 peak is likely a loss of $C_2H_5$. Therefore the molecule is n-butane.

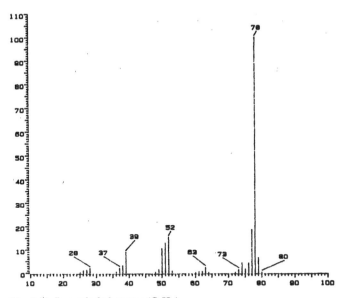

Fig. 3.33. Example 8: benzene ($C_6H_6$).

*Example 8 (Fig. 3.33).* The base peak $m/z$ 78 is clearly the molecular ion. The ratio $(m/z\ 79)/(m/z\ 78)$ is 6.5%, indicating six carbon atoms. The only assignment possible is $C_6H_6$. The $R$ value is 4, suggesting an aromatic ring structure. The lack of extensive fragmentation confirms this structure. The molecule is benzene.

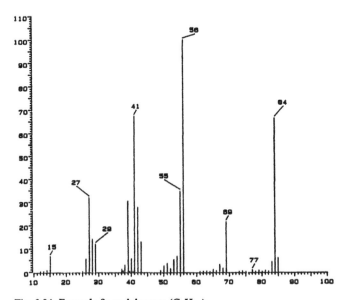

Fig. 3.34. Example 9: cyclohexane ($C_6H_{12}$).

*Example 9 (Fig. 3.34).* This spectrum is very different from example 8, although the structures of the two compounds are similar. The ratio $(m/z\ 85)/(m/z\ 84)$ is 6% which could represent a maximum of six carbon atoms. There are no clusters characteristic of $A+2$ elements evident, so only A elements and possibly oxygen could be included with carbon. Each cluster of peaks is separated by approximately 15 a.m.u., suggesting a cyclic aliphatic hydrocarbon. Other possibilities are alkenes, alkenyl or cycloalkyl carbonyl compounds, cyclic alcohols or ethers. In general, the spectrum is not consistant with the oxygen-containing classes, so we can assume $C_6H_{12}$ as a formula for the molecular ion. The $R$ value is 1, so alkenes or cycloalkane structures are possible. Since the cleavage of a sigma bond in a cycloalkane forms a 1-alkene ion, further interpretation of this spectrum is difficult. However, the large $m/z$ 56 peak, a loss of $C_2H_4$, would favour the assignment of cyclohexane to the structure.

### 3.3.6. Problems

Mass spectra and tabulated mass and abundance values are presented in Figs. 3.35–3.42 as problems in mass spectrum interpretation. The compounds are more complex than the examples interpreted in Figs. 3.26–3.34 but they can be solved by application of the steps outlined in Table 3.5. The compounds in the problem set are examples of the following classes: alkanes, alcohols, ketones, esters, polynuclear aromatics, phenols, and aromatic amines. The solutions to the problems are given in Table 3.6.

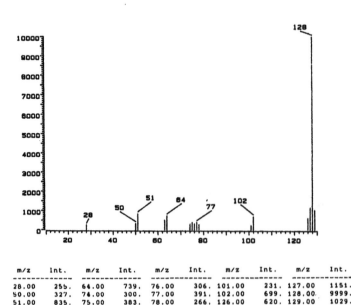

| m/z | Int. | m/z | Int. | m/z | Int. | m/z | Int. | m/z | Int. |
|---|---|---|---|---|---|---|---|---|---|
| 28.00 | 255. | 64.00 | 739. | 76.00 | 306. | 101.00 | 231. | 127.00 | 1151. |
| 50.00 | 327. | 74.00 | 300. | 77.00 | 391. | 102.00 | 699. | 128.00 | 9999. |
| 51.00 | 835. | 75.00 | 383. | 78.00 | 266. | 126.00 | 620. | 129.00 | 1029. |
| 63.00 | 510. | | | | | | | | |

Fig. 3.35. Problem 1: a polynuclear aromatic.

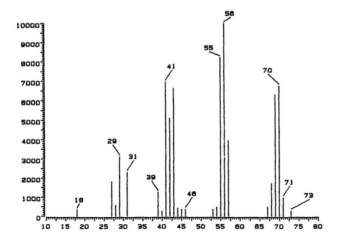

| m/z | Int. | m/z | Int. | m/z | Int. | m/z | Int. | m/z | Int. |
|-----|------|-----|------|-----|------|-----|------|-----|------|
| 18.00 | 330. | 40.00 | 260. | 46.00 | 350. | 67.00 | 460. | 82.00 | 781. |
| 27.00 | 1812. | 41.00 | 6946. | 53.00 | 340. | 68.00 | 1702. | 83.00 | 4034. |
| 28.00 | 561. | 42.00 | 5105. | 54.00 | 470. | 69.00 | 6296. | 84.00 | 5475. |
| 29.00 | 3103. | 43.00 | 6636. | 55.00 | 8207. | 70.00 | 6736. | 85.00 | 400. |
| 31.00 | 2302. | 44.00 | 420. | 56.00 | 9999. | 71.00 | 951. | 97.00 | 360. |
| 39.00 | 1231. | 45.00 | 350. | 57.00 | 3934. | 73.00 | 260. | | |

Fig. 3.36. Problem 2: an alcohol.

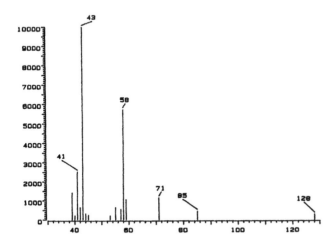

| m/z | Int. | m/z | Int. | m/z | Int. | m/z | Int. | m/z | Int. |
|-----|------|-----|------|-----|------|-----|------|-----|------|
| 39.00 | 1400. | 42.00 | 640. | 45.00 | 240. | 57.00 | 540. | 71.00 | 1140. |
| 40.00 | 210. | 43.00 | 9999. | 53.00 | 210. | 58.00 | 5710. | 85.00 | 440. |
| 41.00 | 2510. | 44.00 | 320. | 55.00 | 640. | 59.00 | 1050. | 128.00 | 274. |

Fig. 3.37. Problem 3: a ketone.

74

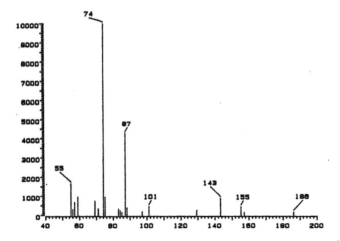

| m/z | Int. | m/z | Int. | m/z | Int. | m/z | Int. | m/z | Int. |
|------|------|------|------|------|------|------|------|------|------|
| 54.95 | 1652. | 69.15 | 735. | 83.15 | 331. | 88.15 | 387. | 143.25 | 873. |
| 56.05 | 299. | 71.15 | 335. | 84.15 | 239. | 97.15 | 181. | 155.25 | 435. |
| 57.05 | 664. | 74.15 | 9999. | 85.15 | 144. | 101.15 | 456. | 157.25 | 144. |
| 58.95 | 949. | 75.15 | 952. | 87.15 | 4265. | 129.25 | 260. | 186.25 | 157. |

Fig. 3.38. Problem 4: an ester.

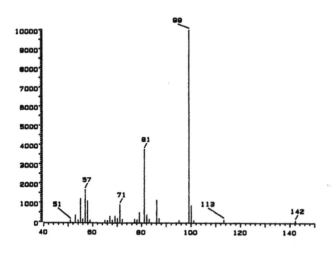

| m/z | Int. | m/z | Int. | m/z | Int. | m/z | Int. | m/z | Int. |
|------|------|------|------|------|------|------|------|------|------|
| 50.95 | 71. | 57.95 | 1061. | 69.15 | 292. | 79.15 | 476. | 95.15 | 60. |
| 52.95 | 335. | 58.95 | 68. | 70.15 | 170. | 81.15 | 3760. | 99.15 | 9999. |
| 54.05 | 77. | 65.05 | 81. | 71.15 | 879. | 82.15 | 363. | 100.15 | 839. |
| 55.05 | 1183. | 66.05 | 54. | 72.15 | 123. | 83.15 | 131. | 101.15 | 59. |
| 55.95 | 117. | 67.05 | 272. | 77.15 | 132. | 86.15 | 1119. | 113.25 | 104. |
| 56.95 | 1656. | 68.05 | 81. | 78.15 | 98. | 87.15 | 164. | 142.25 | 59. |

Fig. 3.39. Problem 5: an alcohol.

75

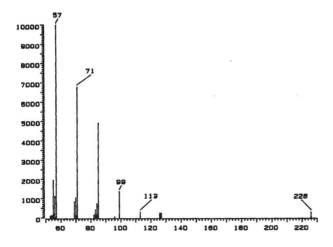

| m/z | Int. | m/z | Int. | m/z | Int. | m/z | Int. | m/z | Int. |
|------|------|------|------|------|------|------|------|------|------|
| 55.00 | 1963. | 69.00 | 826. | 83.00 | 412. | 99.00 | 1344. | 127.00 | 241 |
| 56.00 | 1111. | 70.00 | 1032. | 84.00 | 733. | 113.00 | 297. | 226.00 | 270. |
| 57.00 | 9999. | 71.00 | 6782. | 85.00 | 4912. | 126.00 | 233. | | |

Fig. 3.40. Problem 6: an alkane.

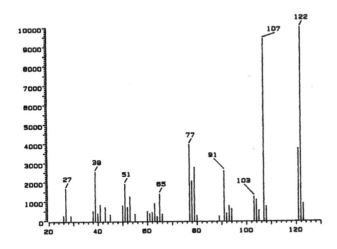

| m/z | Int. | m/z | Int. | m/z | Int. | m/z | Int. | m/z | Int. |
|------|------|------|------|------|------|------|------|------|------|
| 26.00 | 252. | 45.00 | 307. | 62.00 | 423. | 80.00 | 250. | 104.00 | 1070. |
| 27.00 | 1653. | 50.00 | 781. | 63.00 | 886. | 89.00 | 213. | 105.00 | 518. |
| 29.00 | 232. | 51.00 | 1898. | 64.00 | 214. | 91.00 | 2574. | 107.00 | 9397. |
| 38.00 | 493. | 52.00 | 701. | 65.00 | 1378. | 92.00 | 366. | 108.00 | 721 |
| 39.00 | 2530. | 53.00 | 1239. | 66.00 | 329. | 93.00 | 761. | 121.00 | 3741 |
| 40.00 | 372. | 55.00 | 330. | 77.00 | 3933. | 94.00 | 600. | 122.00 | 9999. |
| 41.00 | 823. | 60.00 | 479. | 78.00 | 2053. | 103.00 | 1228. | 123.00 | 872. |
| 43.00 | 702. | 61.00 | 354. | 79.00 | 2755. | | | | |

Fig. 3.41. Problem 7: a phenol.

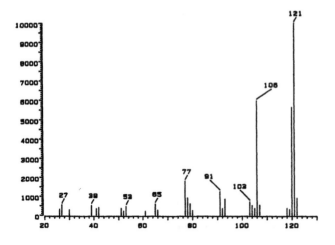

| m/z | Int. | m/z | Int. | m/z | Int. | m/z | Int. | m/z | Int. |
|------|------|------|------|------|-------|--------|-------|--------|-------|
| 26.00 | 319. | 51.00 | 364. | 77.00 | 1786. | 93.00 | 834. | 118.00 | 354. |
| 27.00 | 539. | 52.00 | 206. | 78.00 | 897. | 103.00 | 692. | 119.00 | 299. |
| 30.00 | 280. | 53.00 | 458. | 79.00 | 594. | 104.00 | 506. | 120.00 | 5614. |
| 39.00 | 510. | 61.00 | 209. | 80.00 | 253. | 105.00 | 351. | 121.00 | 9999. |
| 41.00 | 342. | 65.00 | 574. | 91.00 | 1220. | 106.00 | 5948. | 122.00 | 886. |
| 42.00 | 405. | 66.00 | 267. | 92.00 | 346. | 107.00 | 519. | | |

Fig. 3.42. Problem 8: an aromatic amine.

TABLE 3.6

COMPOUNDS ASSOCIATED WITH THE MASS SPECTRA IN THE INTERPRETATION PROBLEMS

| Figure | Compound | Molecular formula | Class |
|--------|----------|-------------------|-------|
| 3.35 | Naphthalene | $C_{10}H_8$ | Polynuclear aromatic |
| 3.36 | 1-Octanol | $C_8H_{18}O$ | Alcohol |
| 3.37 | 2-Octanone | $C_8H_{16}O$ | Ketone |
| 3.38 | Methyl decanoate | $C_{11}H_{22}O_2$ | Ester |
| 3.39 | 1-Propylcyclohexanol | $C_9H_{18}O$ | Alcohol |
| 3.40 | Hexadecane | $C_{16}H_{34}$ | Alkane |
| 3.41 | 2,6-Dimethylphenol | $C_8H_{10}O$ | Phenol |
| 3.42 | 2,6-Dimethylaniline | $C_8H_{11}N$ | Amine |

## 3.4. Suggested reading

1 F.W. McLafferty, *Interpretation of Mass Spectra*, University Science Books, Mill Valley, CA, 3rd ed., 1980.

2 M.C. Hamming and N.G. Foster, *Interpretation of Mass Spectra of Organic Compounds*, Academic Press, New York, 1972.

3 F.W. McLafferty, *Mass Spectral Correlations (Advances in Chemistry Series 40)*, American Chemical Society, Washington, DC, 1963.

4 R.M. Silverstein and G.C. Bassler, *Spectrometric Identification of Organic Compounds*, Wiley, NY, 2nd ed., 1967.

5 V.M. Parikh, *Absorption Spectroscopy of Organic Molecules*, Addison-Wesley, Reading, MA, 1974.
6 K. Feser and W. Köegler, *J. Chromatogr., Sci.*, 17 (1979) 57–63.
7 R.D. Craig, R.H. Bateman, B.N. Green and D.S. Millington, *Phil. Trans. R. Soc. Lond.*, 293 (1979) 135–155.
8 C.J. Porter, J.H. Beynon and T. Ast, *Org. Mass Spectrom.*, 16 (1981) 101–114.
9 H. Budzikiewicz, *Angew. Chem. Int. Ed. Engl.* 20 (1981) 624–637.
10 R.C. Dougherty, *Anal. Chem.*, 53 (1981) 625A-636A.
11 P.E. Miller and M.B. Denton, M.B., *J. Chem. Ed.* 63 (1986) 617–622.
12 D.F. Hunt, G.C. Stafford, F.W. Crow, and J.W. Russell, *Anal. Chem.*, 48 (1976), 2098–2104.
13 G.C. Stafford, *Environ. Health Perspect.*, 36 (1980) 85–91.
14 A.G. Harrison, *Chemical Ionization Mass Spectrometry*, CRC Press, Boca Raton, FL, 1983.

*CHAPTER 4*

# GAS CHROMATOGRAPHY–MASS SPECTROMETRY

Both a gas chromatograph and a mass spectrometer are relatively simple instruments conceptually and the analytical data each produces are easily understood and used. When these two instruments are directly combined into one GC–MS system, the capabilities of that system are not merely the sum of the two instruments; the increase in analytical capabilities is exponential. To realize the potential which lies in the enormous amount of data generated by the GC–MS system a dedicated computer is necessary. Once the computer is in place many techniques of data manipulation become possible to enhance the analytical qualities of the data. To fully appreciate the optimum operation of each component of the system it is

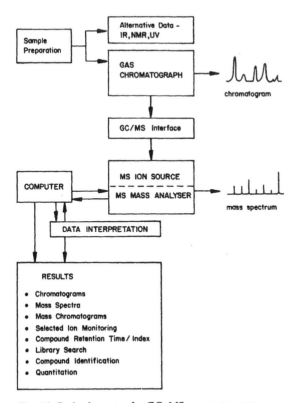

Fig. 4.1. Basic elements of a GC–MS–computer system.

necessary to understand the basic principles by which each functions. The functions provided by the computer that produces the data needed to achieve the analytical results must be well understood to maximize their utility to the analyst. The basic elements of a GC–MS–computer system are indicated in Fig. 4.1.

## 4.1. Vacuum and gas flow

### 4.1.1. Basic principles

Optimizing the performance of a GC–MS instrument depends on knowledge of gas flow and vacuum principles. Chromatographs operate at a wide range of carrier gas flows, temperatures, and quantities of organic compounds per peak. Mass spectrometers have a range of vacuum systems, flow conductances, ion sources, and mass selector designs. The connecting lines throughout the GC–MS system will contain valves, tubings, and orifices, each with individual effects on gaseous flow behaviour. Exact computation of these effects is not possible nor necessary. Approximate calculations can be made which are adequate to determine how these principles influence the operation and analytical results in a GC–MS system. Tables 4.1 and 4.2 summarize the concepts and equations defining these principles.

The mean free path (MFP) is the average distance a gaseous particle will travel before it collides with another particle. Viscous flow occurs when the MFP of molecules is much smaller than the diameters of the orifice of tube through which they are flowing. Molecules then undergo many interactive collisions with each other and move together in a bulk fashion.

Molecular flow occurs when the MFP of molecules is much greater than the dimensions of the tube, orifice, or enclosure involved. Under these conditions

TABLE 4.1

PRINCIPLES OF VACUUM SYSTEMS

| Concept | Description | Equation |
|---------|-------------|----------|
| Mean free path (MFP) | Average distance a particle will travel before it collides with another particle. | $MFP(cm) = 5 \cdot 10^{-3}/P$ ($P$ = pressure in torr) |
| Viscous flow | Occurs when MFP is much smaller than dimensions of an enclosure. | $P \cdot D > 0.5$ ($P$ = pressure in torr, $D$ = diameter in cm) |
| Molecular flow | Occurs when the MFP is much greater than dimensions of an enclosure. | $P \cdot D < 0.5$ |
| Conductance (C) | Expresses gas flow through an orifice or conduit. | $Q = C(P_1 - P_2)$ ($Q$ = quantity in, $C$ = liter/s, $P_1 - P_2$ = pressure drop in torr) |
| Speed of vacuum pump (S) | Measures of the ability of a vacuum pump to move gas across a plane in a system. | $S = Q/P$ |

TABLE 4.2

CALCULATION OF CONDUCTANCES

| Concept | Description | Equation |
|---------|-------------|----------|
| Combined conductances | Total conductance ($C_T$) calculated using equations for series and parallel configurations | *Parallel* <br> $C_T = C_1 + C_2 + C_3$ <br> *Series* <br> $\dfrac{1}{C_T} = \dfrac{1}{C_1} + \dfrac{1}{C_2} + \dfrac{1}{C_3}$ |
| Viscous flow conductance | Calculations of conductance through a conduit | $C = \dfrac{3.3 \cdot 10^{-8}}{\eta \cdot L} \cdot D^4 \cdot P$ <br> ($D$ = diameter in cm, $\eta$ = viscosity in poise, $L$ = length in cm) |
| Molecular flow conductance | Different equations used for orifices and conduits | *Orifice* <br> $C = 9.1 \cdot D^2 \, (28.7T/293M)^{1/2}$ <br> *Conduit* <br> $C = \dfrac{12.1\, D^3 \, (28.7T/293M)^{1/2}}{L}$ <br> ($T$ = temperature in K, $M$ = molecular weight) |

molecules collide with the enclosure walls and not with each other. They then move independently at velocities controlled by their molecular weights and temperatures. The approximate equations describing this behaviour in Table 4.2 are correct to within 10%.

Conductance ($C$) is a term used to express gas flow which passes through an orifice or conduit. Under molecular flow conditions conductance can never exceed the volume throughput permitted by the thermal velocity of the molecules and the area of the conducting orifice or tube. This can be illustrated with the simple example shown in Fig. 4.2. The volume of gas which will pass through a 6 cm² orifice can never exceed 60 liter/s, regardless of how much capacity the vacuum pump has. This is calculated as a product of molecular thermal velocity ($10^4$ cm/s) and orifice area (6 cm²). This indicates that small orifices and narrow diameter conduits connecting vacuum pumps to GC–MS systems seriously reduce the pump capacity.

The total conductance of a system can be calculated from the individual conductances of each component. Conductances in parallel are additive, but those in series are calculated by the inverse equations shown in Table 4.2.

Another concept needed for analysis of vacuum systems is the speed of a vacuum pump. The defining equation resembles that for conductance. Although pumping speed ($S$) and conductances may be defined in the same units and may be numerically equal, they are never equivalent in meaning. Conductance implies a pressure gradient and is a geometrical property. Pumping speed is applied to any

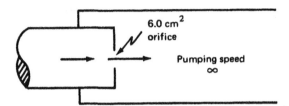

Under molecular flow conditions, conductance can never exceed volume throughput permitted by thermal velocity of molecules and area of conducting orifice.

Fig. 4.2. Maximum conductance of an orifice.

plane in the system which can be considered a pump with an ability to remove gas. In a mechanical pump an eccentric cylinder sweeps out a volume of gas with each rotation. Its speed is the product of that volume and the rotational frequency.

### 4.1.2. Analysis of vacuum and gas flow

The problem in a GC–MS system is to place as much as possible of the organic compound in a GC peak into the MS ion source without exceeding the vacuum requirements in any part of the mass spectrometer. In the MS ion source the pressure must be kept below $10^{-4}$ Torr to avoid ion-molecule reactions caused by a short mean free path and leading to unrecognizable fragmentation patterns. In the mass analyzer section of the mass spectrometer the MFP must be greater than 200 cm, which corresponds to a pressure of $10^{-5}$ Torr, so that scattering of the ion beam which results in degredation of performance is avoided.

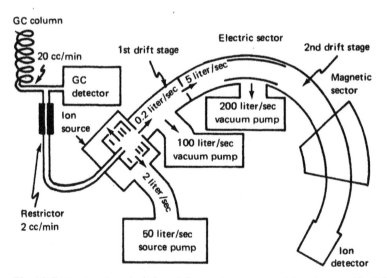

Fig. 4.3. Parameters for calculation of flow and vacuum conditions in a specific GC–MS system.

TABLE 4.3

ANALYSIS OF FLOW AND VACUUM CONDITIONS IN GC–MS SYSTEM OF FIG. 4.3

|  | $Q_{entering}$ | $C_{out}$ | P(torr) | MFP (cm) |
|---|---|---|---|---|
| Ion chamber | $2.5 \cdot 10^{-2}$ | 2.2 | $1.1 \cdot 10^{-2}$ | 0.5 |
| Source pump | $2.2 \cdot 10^{-2}$ | 50 | $4.4 \cdot 10^{-4}$ | 10 |
| Mass analyzer | $2.2 \cdot 10^{-3}$ | 100 | $2.2 \cdot 10^{-5}$ | 200 |

Calculations can be made on a specific GC–MS system shown schematically in Fig. 4.3 to illustrate use of the equations and concepts. Using the equation $P = Q/C$ and $P = Q/S$ the analysis of conditions listed in Table 4.3 can be made.

These calculations show that the source pressure is somewhat high. The source vacuum pump is operating near its limit (MFP is approximately 10 cm) and does not have capacity for much more GC effluent. It can be seen that without the separate source vacuum pump the GC effluent would have to be removed by the 100 liter/s mass analyzer vacuum pump. Were this to be done, then only 1/10 the amount of GC effluent could be admitted and still maintain the low pressure and 200 cm MFP needed.

### 4.1.3. Interfaces

A special interface between the gas chromatograph and the mass spectrometer is needed because the gas chromatograph operates at atmospheric pressure and the mass spectrometer ion source at $10^{-5}$ Torr. In the gas chromatograph sample molecules mixed with a carrier gas. This means the GC peak must go through a pressure difference of $10^8$ fold. The carrier gas in a GC peak must be removed to avoid destroying the high vacuum conditions. This is accomplished by an interface which preferentially removes the carrier gas molecules and transfers the GC peak components to the MS ion source.

In some cases, such as wall-coated open tubular (WCOT) column chromatography with its low flow rates of 1–3 ml/min, the required pressure reduction can be accommodated by the vacuum system of the mass spectrometer. Therefore, a direct interface may be used. Both carrier gas and sample pass into the ion source of the mass spectrometer where the carrier gas is pumped away at a much more rapid rate than the sample molecules. The benefit of a direct interface is that the operator is assured that all of the sample passes from the gas chromatograph into the mass spectrometer. A direct interface cannot be used with packed columns because of the higher carrier gas flow rates (15–40 ml/min). Fig. 4.4 shows the designs of various interfaces that have been developed and used. Some of these interfaces are of historical interest only since the advent of wide-spread use of WCOT columns.

*Effusive interface.* Operation of the effusive interface uses the principle of separation of carrier gas and sample molecules based on the difference in mass. The effluent GC peak passes into a tube of porous glass enclosed in a vacuum. As the molecules and carrier gas enter the porous tube, the light helium carrier gas will preferentially pass through the fritted glass into the vacuum region and be pumped

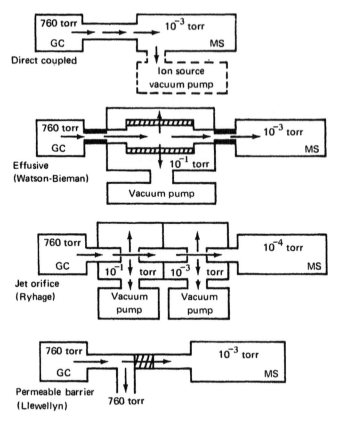

Fig. 4.4. Design of GC–MS interfaces.

away. The ability of a molecule to pass through small pores in the fritted glass is inversely proportional to its mass. Therefore, few sample molecules for high molecular weight compounds will be lost. This is not true for compounds of low molecular weight where there is a relatively small difference in mass between the carrier gas and sample. This is one of the limiting factors of an effusive interface that has reduced its use in practice. It also has a large dead volume and surface area, factors which create GC peak broadening, so that it works poorly with packed columns and not at all with WCOT columns.

*Jet orifice interface.* The jet orifice interface is based on the principle of passing the sample and carrier gas from the exit of a GC column through a small jet orifice. In passing through the jet, the molecules increase in velocity and pass into a vacuum region in the direction of a collector orifice directly lined up with the first orifice and connected to the MS ion source. The two orifices are only about 0.5 mm apart. The light carrier gas (He, $N_2$, $H_2$) is preferentially pumped away in the vacuum region. The heavier molecules with their greater momentum are more resistant to changing direction and go directly into the orifice on the opposite side.

The efficiency of a jet interface is dependent upon carrier gas flow and must be designed for optimum flow rates. Jet interfaces are made of glass for sample inertness, and are applicable to a wide range of sample types. It functions well with both packed and WCOT columns.

*Permeable membrane interface.* The permeable membrane interface is based on the principle of passing a sample and carrier gas from the gas chromatograph over the surface of a silicone rubber membrane with the ion source vacuum chamber on the opposite side of the membrane. The organic molecules dissolve on the surface of the membrane and permeate through into the vacuum region of the mass spectrometer. The non-organic carrier gas molecules do not dissolve and are vented to atmosphere.

The membrane interface has a very high enrichment factor due to the nature of its design, however, its yield is lower than in other interfaces. Since its operation is based on a principle of solubility of the organic molecule onto the membrane surface, high molecular weight, non-volatile, and polar molecules are discriminated against due to their lesser permeability in the membrane.

The permeable membrane interface is easy to use, inexpensive, and causes few problems in operation. It does not become clogged as does a jet orifice separator. The membrane interface is preferred for general types of organic analyses and packed column chromatography. With some modification, such as adding a make-up gas, it can be used with WCOT columns but it is not preferred.

*WCOT column interfaces.* WCOT column GC–MS interfaces have the same requirements as packed columns of chemical inertness and pressure reduction from the column end to the high vacuum in the ion source. Additional parameters must be addressed, however, due to the low flow rate used, the pressure changes in the interface when temperature programming and the effect of dead volume on the chromatographic resolution. The recent development of fused silica WCOT columns which are more chemically inert make these parameters even more pronounced.

The low flow rates by WCOT columns provide a relatively easy solution to the pressure-vacuum problem. If the pumping system on the mass spectrometer is adequate, the entire effluent may pass into the source and we have a direct interface. This provides for the most dependable chromatography except for the effect on GC retention caused by the end of the column being exposed to vacuum. This configuration can be difficult to work with. For example columns may not be removed or changed without venting the mass spectrometer. In addition the mass spectrometer cannot be isolated from the gas chromatograph during use.

Low flow rates and narrow peak widths also emphasize the need for carefully made connections between the column and the interface. Dead volumes and active metal surfaces show a pronounced effect on peak tailing and GC peak intensity. Interfaces have been developed which attempt to minimize these effects.

*Direct split interface.* No sample enrichment is performed in this interface. The effluent is split in the interface to accommodate the pumping capacity of the mass spectrometer. Typically the portion which is split is vented to atmosphere thereby maintaining the end of the column at atmospheric pressure. The split ratio will vary during temperature programming due to gas viscosity changes and therefore this

Fig. 4.5. Design of the open-split interface for WCOT columns.

interface is not well suited for certain types of quantitative measurements. To minimize dead volume, column-interface connections are frequently made by fusing narrow-bore platinum tubing to the glass column. Platinum is also inert to most organic molecules.

*Open split interface.* The development of fused silica columns, the need to change columns regularly and the changing gas dynamics at the interface have led to the popularity of the open split interface. In this case the end of the column is butted against a restrictor line going to the MS ion source. A sleeve surrounds the butted tubes and a small flow of helium is passed around the sleeve.

Columns may be easily changed because the restrictor maintains vacuum in the source. Connection is low dead volume and flows can be changed with no effect on the MS conditions such as during temperature programming. The open split interface is the easiest to use and gives the best chromatographic integrity short of the direct interface. It is shown in Fig. 4.5.

*Comparative evaluation of interfaces.* Because of the difficulty of measuring the efficiency ($N$) and yield ($Y$) very few data are available to analyze the performance of interfaces. Table 4.4 summarizes some available information.

TABLE 4.4

COMPARATIVE PERFORMANCE OF GC–MS INTERFACES

$N$ = (concentration in MS source)/(concentration in GC peak). Yield = (quantity in MS source)/(quantity in GC peak). $H$ = (peak width in MS)/(peak width from GC).

| Classification | Efficiency, $N$ | Yield (%) | Delay (s) | $H$ | Inert | Carrier gases |
|---|---|---|---|---|---|---|
| Perfect | Infinite | 100 | 0 | 1 | Yes | All |
| Jet | 100 | 40 | 1 | 1–2 | Yes | He, $H_2$ |
| Direct | 1 | 1–100* | 1 | 1 | Yes | All |
| Effusive | 100 | 50 | 1 | 1–2 | Maybe | He, $H_2$ |
| Permeable membrane | 1000 | 80–95 | Variable | 3 | Maybe | Inorganic |

* Equals split ratio

## 4.2. Computerization

Although a gas chromatograph can be used to obtain good data without the use of a computer, such is not true of a GC–MS system. The amount of useful data generated is so overwhelming that it is neither practical nor possible to obtain and analyze it by standard methods. If a hypothetical case were visualized in which a person would record and process by hand the data generated during a GC–MS run, an estimate of the time involved would be at least a factor of 400 to 1 over a computerized system. That means it would take more than one year to do what a computerized GC–MS system does in one day.

### 4.2.1. Computerized operation

The advantages of computerization for GC instrumentation have been discussed previously. These comments still apply to GC–MS computerization. However, the much greater complexity of GC–MS instruments combined with their fast scan rates and huge data generating potential make computers a mandatory requirement for the GC–MS systems. Most GC–MS instruments are already equipped with computer and data system when initially purchased. For those instruments not so equipped there are data systems available that can be purchased separately and interfaced to either magnetic sector or quadrupole analyzers. Above, an example was presented of the large amount of data which can be generated during a single GC–MS analysis. To further illustrate the multi-task requirements of the computer, a more detailed example is presented here.

A sample of diesel engine particulate matter is collected for characterization of the organic compounds present. After extracting the sample in a suitable solvent and concentrating this extract by a factor of 1000, it is analyzed by GC and is found to be a complex mixture of over 200 compounds at widely varying concentrations. It is desired to identify as many of these compounds as possible and quantify any carcinogenic or mutagenic substances that are detected. The GC separation required a 90-min WCOT column temperature programmed run.

*Instrument initialization.* Before GC–MS analysis, it is necessary to calibrate the mass analyzer. This is performed by scanning a calibration compound for which the peaks are well known from previous work. Common calibrating compounds are perfluorotri-*n*-butylamine and perfluorokerosine. A calibration table is generated and stored by the computer so that individual $m/z$ values can be determined during scanning operation of the analyzer. In some instruments the entire tuning process is performed by computer control, including a diagnostic report which indicates if the system is operating within specifications. In others, the actual tuning is performed manually, although the generation of the calibration table is controlled by the computer.

Analysis conditions can now be entered into a computer file. These include initial GC temperature, injection port temperature, program rate, final temperature, and mass spectrometer conditions. Most data systems allow the user to store several analysis conditions files so that all GC–MS conditions for specific types of routine samples can be entered by specifying a single file number. When all conditions are

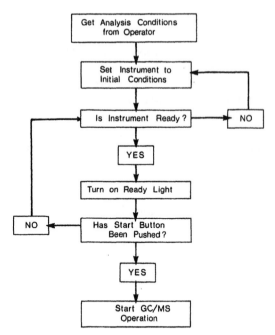

Fig. 4.6. Simplified flow chart of initial procedures for computerized GC–MS instrument.

at initial values, an indication is given to the analyst who may then inject the sample. Fig. 4.6 shows a simplified flow chart of these initial procedures.

*Instrument control and data acquisition.* Upon injection, several operations are performed under computer control. Control of the gas chromatograph is done by comparing the measured oven temperature to the calculated correct value at that time. Mass spectrometer control includes initiating the mass scans, converting numbers from the detector into mass, intensity data, and sending these data onto a storage device such as a disk. Fast scan rates (0.5–2 s) must be achieved for the narrow peaks obtained for WCOT GC. A typical run-time sequence might be as illustrated in Fig. 4.7. The MS diagnostic procedures may include checking to ensure the ion source filament is operating and that the analyzer vacuum is low enough. The GC oven temperature is monitored continually to ensure it is at the proper value. If the temperature is too low, a heater is turned on, and if the temperature is too high, cooling can be effected by opening the oven door or delivering liquid into the oven as coolant. Before storing data, the storage device must be examined to ensure enough free space remains there to accept the new data. At any time during the analysis, the operator can interrupt the program to change certain parameters such as temperature program rate or total analysis time.

Operations illustrated in Fig. 4.7 are repeated rapidly enough to obtain and store a complete mass spectrum every 2 s. At the completion of the 90-min WCOT column analysis of the diesel extract, 2700 mass spectra will be stored. Each peak in a single mass spectrum consists of a mass and abundance value, so a typical mass

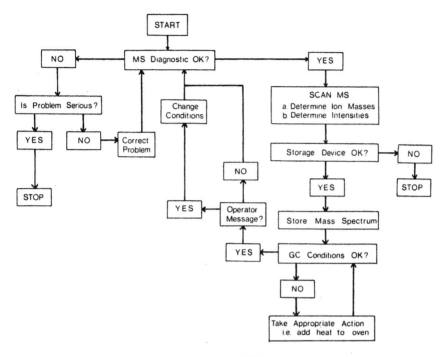

Fig. 4.7. Runtime sequence of computer controlled GC–MS analysis.

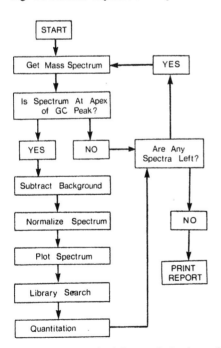

Fig. 4.8. Computerized data analysis of stored mass spectra.

spectrum of about 150 peaks requires the storage of 300 numbers. For the entire analysis, storage space for more than $300 \times 2700 = 810,000$ numbers is required. The actual storage required is greater since other data such as retention times must be stored. Although much of this information can be discarded, for example, scans taken during times in the analysis where no GC peaks were detected, the task of processing such a vast amount of data is too labour intensive to perform manually for even a single analysis.

*Post-run data processing.* The value of the computer has been established without even considering the processing of the vast amount of data which is stored. Most of the tasks described above are transparent to the user. A large portion of the tasks of the computer has been accomplished without requiring any interaction with the operator. After the instrumental analysis, a large number of data processing operations can be performed. Many of these data analysis techniques are described in section 4.3. Methods of identifying mass spectra by comparing with known spectra in reference files are described in section 4.4.

For our 2700 stored mass spectra of the diesel extract analysis, typical data reduction procedures could include those illustrated in Fig. 4.8. If any of the identified compounds are carcinogenic or mutagenic, their peak areas can be obtained from the stored data. By analyzing known amounts of these substances using the same GC–MS conditions, quantitation of these compounds can be performed. The final report can include information such as the probable identity, peak area and retention index for each substance identified. Compounds not identified by simple library search may need manual interpretation, although other computer techniques have been developed to assist this task. Some of these are described in section 4.4.

### 4.2.2. Characteristics

*Speed of operation.* Many operations must be performed during a GC–MS analysis, some of which were illustrated in Fig. 4.7. These must be executed fast enough to allow most of the computer's time to be devoted to obtaining and storing mass spectrometer data. With instruction execution times on the order of microseconds, modern computing devices can easily accomplish the necessary tasks. However, the design of GC–MS software is complex and all commercial data systems do not possess the same capabilities. With efficient program design, some software systems allow the simultaneous operation of more than one GC–MS analyzer. Some instruments use more than one computer, where the principal device is a minicomputer that interacts with the operator and controls all data processing, and a microcomputer is used for instrument control during data acquisition. The principal advantages of this configuration are increased speed of operation and use of the minicomputer to process data at the same time as it is being acquired. This type of operation can be achieved solely by software design, but is more efficient with the dual processor system.

An important consideration is the speed at which data are converted from voltages to numbers (A/D conversion rate). The computer does not perform this function directly, but must process the numbers generated by an A/D converter.

For GC–MS operation, these devices must be operated at frequencies of 50–100 kHz. Several data points are sampled across each mass peak to obtain accurate mass information. Some preliminary processing is required before ion mass and abundance information is transmitted to a mass storage device. All data transfer is performed by computer control using a transmission system called a data bus.

*Storage devices.* All modern GC–MS data systems use disk drives for data storage. A great deal of storage area is required for such samples as the diesel particulate extract described previously. A typical single platter hard disk can store 5–10 such analyses. Most systems are equipped with a dual disk drive for which data are normally stored on a removable disk while operating programs and analysis conditions files are on a fixed disk. Multiple-platter drives in a sealed unit (Winchester disk drive) are available for very large data storage. Information on these devices is transferred to tape for long-term (archival) storage. Tape units for storing data can be purchased separately and used for any data system. Removable disk platters and archival tape storage provide effectively limitless data storage. Smaller, inexpensive disk drives, called floppy disks, have been used on some systems due to their greatly reduced cost and easy maintenance compared to hard disk drives. Because of their limited storage, compromises have to be made during acquisition of data. The entire diesel analysis described earlier would require 3–4 of these floppy disks. By storing only scans taken at the apex of peaks, the entire analysis could be stored on a single floppy disk, although post-run data processing options will be limited.

*Other peripheral devices.* Computerized systems generally have available other devices in addition to disk drives. The most common is a video terminal display through which commands are entered and data displayed. A printer/plotter is necessary to obtain a printed record of the analysis. Some systems can handle several disk drives, terminals, plotter, printer, and tape drive. Not all data systems have this flexibility, which depends upon the specific computer's hardware capabilities and available support software. Because it may not be possible to add certain peripheral devices to a specific computer after the initial installation of the system, future expansion plans should be considered when setting up a GC–MS–computer system. An inexpensive device which can be valuable for many systems is an acoustic coupler or modem, which allows the user's computer to communicate with remote computers via telephone lines. Using this device, mass spectral identification can be performed using sophisticated search methods on remote computers for a nominal charge. Such systems are described in section 4.4.

*Interfacing.* An important part of computerized operation is the electronic interfacing between the computer and other devices. Advanced concepts are necessary to ensure the proper priority for operations, since the computer can only execute one instruction at a time. Fortunately, the user seldom has to be concerned with this aspect of GC–MS–computer operation, especially with complete, prepackaged systems. Many peripheral devices are "intelligent", possessing their own specially programmed computer chips, which allows the computer to spend more time performing other duties. Details of computer interfacing principles are complex, however, it is not necessary to understand them to achieve effective GC–MS operation. It is important to realize that limitations exist in the computer's capabilities, and the same serious consideration should be given to the data system and associated peripherals that is given to the GC–MS analyzer when choosing a system.

*Software support.* Although programs for instrument operation and GC–MS data analysis have lagged behind instrument hardware development, many instrument purchases are now being made on the basis of available software, since hardware features on competitive systems are often similar. Consequently, the power of data systems has increased greatly. In addition to the routine acquisition and analysis procedures already described, useful programs include text editors to modify or create data or program files, molecular weight and formula calculators, diagnostic programs, and file management programs to list or transfer or modify disk files. Advanced systems also have one or more high level languages to enable users to program their own specific applications software. Many such user-developed programs have been reported in the technical literature. An important aspect of software development is that GC–MS systems can be upgraded by software improvements instead of by purchasing expensive hardware.

## 4.3. Data analysis

Modern GC–MS systems acquire data by continuous repetitive scanning of the GC column eluate. The rate of scanning is predetermined by the operator; usually values in the range 0.1–1.5 s per scan are used. Each scan results in a series of $m/z$ values and abundances stored in computer memory or associated devices such as a disk drive. The data system of a GC–MS unit is capable of displaying the stored data in a number of ways to enable the analyst to follow the course of the GC run as it progresses, and to analyze the data acquired.

### 4.3.1. Reconstructed gas chromatogram

It is desirable to monitor the chromatographic run during a GC–MS analysis to insure proper operation of the system. A chromatogram can be reconstructed (RGC) as components elute from the column by a plot of the total ion current stored and summed by the computer for each spectral scan. This record is similar to the chromatogram of the GC effluent produced by a flame ionization detector (FID) and can be used to identify the mass spectrum associated with a specific chromatographic peak. Since the mass spectrometer and FID respond differently to compounds, the relative sensitivities of the GC peaks will differ between the two traces.

The RGC is also designated by TI (total ions), TIC (total ion current), and RIC (reconstructed ion current). It is generated during the analysis by summing the ion currents of all ions in each mass spectrum and plotting these values as a function of mass spectrum scan number. After the run the RGC can be plotted in its entirety from data stored by the computer using both the mass spectral scan number and its related retention time. Most data systems enable any chosen section of the RGC to be displayed on a CRT screen to enlarge the details for closer examination.

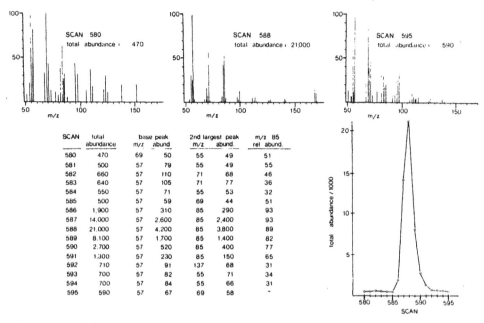

Fig. 4.9. Example illustrating the generation of the RGC by plotting total abundance of ions in stored mass spectra.

Fig. 4.9 shows how the RGC is generated. Scan data are presented for sixteen consecutive mass scans obtained during the analysis of a real environmental sample. Three of the actual mass spectra obtained are also shown. By plotting the total ion abundances as a function of scan number, the RGC trace was generated. The peak shape is well defined even though it is delineated by only six scans (586–591). Scans 580–585 and 592–595 are representative of general background. Two of the background scans plotted (scans 580 and 595) do not have a well defined mass spectral pattern as does scan 588, which was taken at the top of the GC peak. A clear indication of this is that scan 588 contains a small isotope peak after each principal $m/z$ value, while no such isotope peaks are observed in scans 580 and 595.

The shape of the peak could also be followed by plotting the abundances of the principal ions in the mass spectrum of the eluting substance. Data for the two most important peaks are shown in Fig. 4.9. Although $m/z$ 57 appears as the base peak in almost every scan listed, its abundance changes in a more-or-less random fashion until scan 586, where the $m/z$ 57 abundance increases rapidly to the apex of the GC peak, then decreases until the peak has completely eluted. The second largest peak in the mass spectra of scans 580–595 consists of a variety of $m/z$ values except during the elution of the peak, where $m/z$ 85 is always the second most abundant ion. It is important to note that the relative abundance of $m/z$ 85 compared to the most abundant ion ($m/z$ 57) is not constant during the elution of the GC peak. The ideal ratio should be observed in the scan taken at the apex of the GC peak.

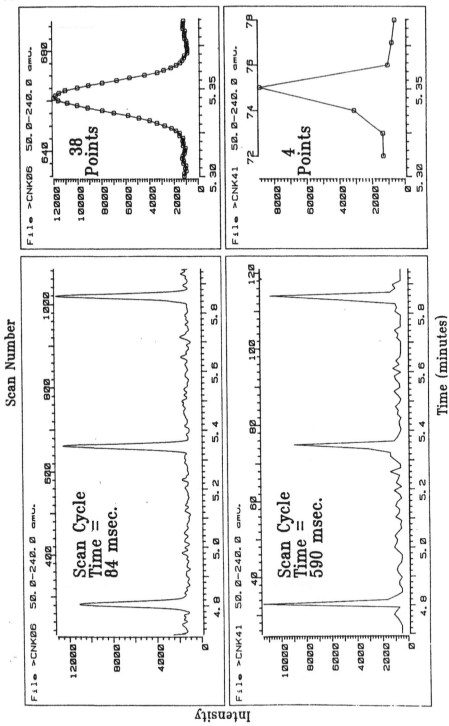

Fig. 4.10. Effect of mass spectrum scan speed on RGC plot.

Differences in mass spectral patterns observed on the leading and trailing edges of GC peaks can be great enough to affect identification by computer search systems.

The RGC sometimes fails to show resolution between two overlapping GC peaks because only one point is produced for each mass scan and a truly continuous plot is not possible. The fewer scans taken during elution of a GC peak, the less faithfully it will be reproduced. Since some GC peaks from WCOT columns may be only 5–10 s wide, the mass spectrometer must be able to scan rapidly (about 0.1 s per scan) to accurately reproduce the GC peak. This is illustrated in Fig. 4.10. In most cases, a scan speed of 1.0 s per scan is fast enough, even during WCOT column operation, to provide a reasonably accurate RGC trace.

There is another aspect in producing GC–MS data for which the scan speed is very important. The mass spectral scans taken during elution of a GC peak must be sufficiently fast that the concentration of the GC peak component does not change enough to distort the normalized mass spectrum. For example, if a GC peak is 10 s wide and two 5-s mass scans were taken during its elution, one mass spectrum would have higher relative abundances for high mass ions. This would be entirely due to the changing concentrations on the front and rear slopes of the GC peak. The two normalized mass spectra resulting would not be recognizable as due to the

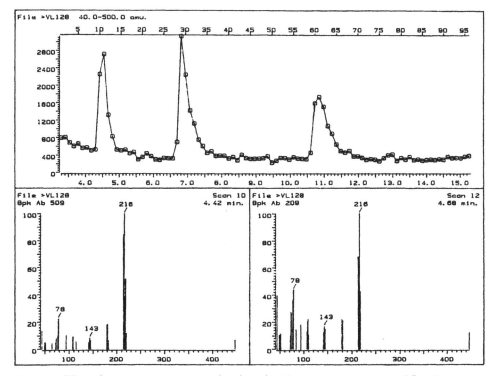

Fig. 4.11. Effect of mass spectrum scan speed on integrity of mass spectrum across a GC peak.

same compound. The mass spectra shown in Fig. 4.11 indicate how a slow scan speed across a 3-s GC peak affects the mass spectral pattern.

### 4.3.2. Mass chromatogram

An RGC plot shows all the GC peaks eluted but gives no mass spectral information. The information present in the stored mass scans can provide valuable diagnostic data for identification of GC peak components. Mass chromatograms (MCs) can also be plotted in which the ion abundances of only a few selected $m/z$ values are extracted from each stored mass spectrum and plotted versus the spectrum number. As in an RGC, the spectrum numbers can be related to retention times.

The advantage of this technique is that by judicious choice of the $m/z$ value which is plotted, selectivity is achieved for the analysis of specific compounds or compound classes. For example, in the analysis of normal alkane hydrocarbons the $m/z$ value 85, a prominent ion in the mass spectra of $n$-alkanes, is usually chosen. The only peaks appearing in the MC of $m/z$ 85 will be from those compounds having a mass 85 ion in their mass spectra. Other mass spectra will not contribute to the MC plot. If a small hydrocarbon peak overlaps with a large peak of a compound that does not have $m/z$ 85 in its mass spectrum, only the hydrocarbon peak will be observed in the MC. A molecular ion MC of those compounds having definitive molecular ions can be very useful in indicating their presence and identifying the GC peaks associated with them.

The use of MCs can be very effective for identification if the proper ions are chosen. Obtaining a peak at the correct mass and retention time for a specific compound is strong evidence for the presence of that compound in the sample, even if a recognizable mass spectrum of the substance is not obtained. If all of the mass spectra in an analysis are stored, an MC can be generated for every $m/z$ value within the scan limits of the mass spectrometer.

The MC is especially useful for rapid surveys of complex GC–MS data to determine the presence and retention times of compounds which contain a given ion mass. This clarification of data simplifies the search for compounds which exhibit a specific, characteristic ion mass.

Fig. 4.9 illustrates how an MC is generated. Instead of plotting the total scan abundances, only the abundance of a specific $m/z$ value in each scan is plotted in the same manner as the RGC. By plotting either the $m/z$ 57 or $m/z$ 85 abundances from Fig. 4.9, the same peak shape would be observed. Although the peak height would be lower by plotting only one selected ion, the background noise would also be reduced.

Fig. 4.12 is the complete RGC for the sample analysis from which the example shown in Fig. 4.9 was taken. The peak at scan 588 is only a minor component of the entire mixture. Also plotted in the figure is a mass chromatogram of $m/z$ 360. This $m/z$ value is characteristic of a group of isomeric compounds known to be present in the sample. Even though these compounds are barely observed as minor components of the RGC trace, they are easily distinguished in the MC plot of $m/z$ 360.

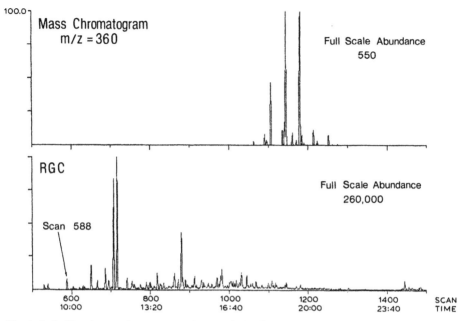

Fig. 4.12. RGC and mass chromatogram from analysis of a real sample.

Note the difference in the full scale abundance values in the RGC (260,000) and MC (550) traces.

### 4.3.3. Selected ion monitoring

Selected ion monitoring (SIM) is used for the detection of very low ion abundances. In this technique, the mass spectrometer is only tuned to a few pre-selected ions; complete mass spectra are not taken during the GC–MS analysis. The gain in sensitivity over scanning the complete mass spectrum for a specific ion is proportional to the ratio of the time the MS is tuned to that ion, to the time that the MS spends on the same ion when scanning to obtain a complete mass spectrum. Consider a mass spectrum covering a 500 a.m.u. range that was scanned in 2 s. The mass spectrometer would only be tuned to each specific $m/z$ for about 2/500 s. If the mass spectrometer was tuned to only one ion for the entire 2 s, this would represent an increased sensitivity factor of 500 for that ion. The SIM plot looks identical to that of an MC but at much higher sensitivity.

In practice, a few ions (6–10) are sequentially monitored. Increased sensitivities of 50–500 for specific compounds are obtained in SIM, depending upon the instrumental conditions and number of ions monitored. Another advantage of SIM is increased selectivity, since ions other than the few chosen will not be detected. However, a poor choice of ions to be monitored can only be corrected by re-analyzing the sample. This is in contrast to the use of MC, which can be generated for every $m/z$ value in each scan stored by the computer.

The term mass fragmentometry is often used in place of SIM. Since molecular

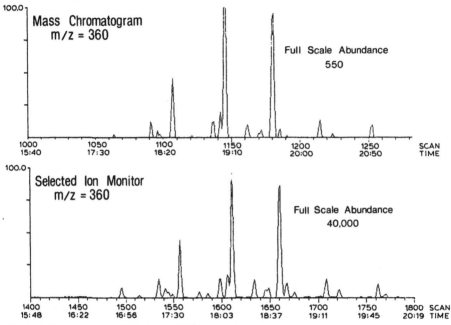

Fig. 4.13. Comparison of SIM plot to MC for a real environmental sample.

ions of compounds may be monitored, not just fragment ions, the term SIM is the more applicable designation. Fig. 4.13 shows the results of a SIM analysis of the same sample illustrated in Fig. 4.12. Twelve ions were used for the SIM analysis, including $m/z$ 360. The mass chromatogram and SIM analysis of $m/z$ 360 have identical patterns, however, the abundances of peaks in the SIM plot are almost 100 times greater. Several minor peaks are also detected in the SIM analysis that were not observed in the MC plot. This indicates that for ultra-trace analysis where the lowest detection limits are needed, SIM must be used. Amounts lower than 1 pg can be detected on some systems, however, optimization of all GC–MS parameters is required.

### 4.3.4. Background subtraction

When plotting the RGC, a significant background component is often observed. This may be due to column bleed, electronic noise, or other factors. If the general level of background ions is large, the overall chromatographic pattern will be difficult to observe, and small peaks may be completely obscured. By re-calculating the baseline of peaks across the RGC, the general background can be removed, allowing individual peaks in the RGC to be more easily observed.

This form of background subtraction is observed in Fig. 4.14. The top trace is an RGC trace of a GC–MS analysis. Abundances are normalized to the largest as 100%. Scan numbers and retention times (min:s) are shown on the abscissa. A large background component is present in this plot. By using a background subtraction

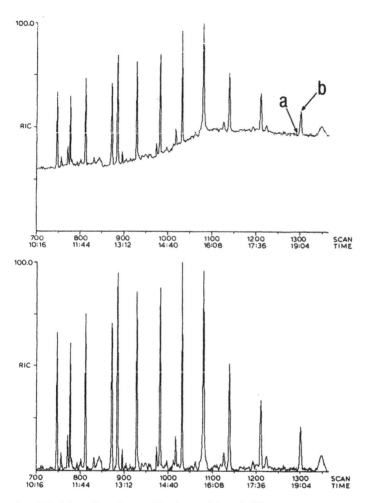

Fig. 4.14. Subtraction of general background from RGC trace.

program of the INCOS data system, the bottom trace of Fig. 4.14 was generated. Peaks are now displayed using the full available plotting range. Note that the full scale abundance of the top trace is 75648, while that of the bottom trace is 41728. The subtraction was performed by giving a single command to the data system. The original unsubtracted data are still stored in the computer, and can be re-plotted at any time.

Individual mass spectra taken during the GC–MS analysis shown in Fig. 4.14 all contain a significant contribution from background ions. Mass spectra are seldom free of background, however, a large background will distort the mass spectra of minor components and make identification difficult or impossible. In many cases, cleaned up mass spectra can be obtained by subtracting another mass spectrum which contains only background ions. This technique of background subtraction is

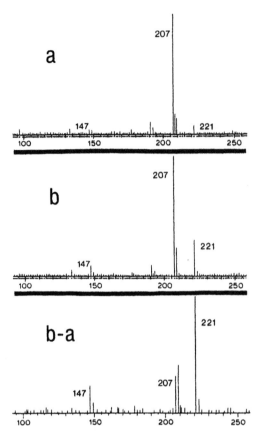

Fig. 4.15. Illustration of background subtraction of mass spectrum. A background spectrum (a) is subtracted from the mass spectrum of an unknown peak (b) to produce the corrected spectrum (b–a).

often successful since many of the background ions are common to a large number of scans, while ions due to the sample component will have maximum abundance only at the apex of the GC peak. The simplest type of subtraction involves a mass-by-mass subtraction of ion abundances in the mass spectrum at the base of a peak from abundances of ions at corresponding masses from the mass spectrum at the apex of the peak. Improved results can sometimes be obtained by summation of two or three spectra taken across the top of a GC peak, and subtraction of several spectra from both the leading and trailing edges of the peak. This method is required to resolve overlapping peaks.

Fig. 4.15 illustrates the advantages of background subtraction. The top mass spectrum (a) was obtained just before elution of a minor component of Fig. 4.14. In the mass range shown ($m/z$ 100–250) there is a background contribution at almost every $m/z$ value. The next spectrum shown (b) is from the apex of the GC peak. The background peak ($m/z$ 207) is still the base ion, while principal ions from the eluting substance at $m/z$ 147 and 221 appear as minor, compared to $m/z$ 207. By subtracting these two mass spectra, the corrected spectrum $(b - a)$ is obtained. The base ion of the eluting component is shown to be at $m/z$ 221. Many of the minor components have been removed completely. A contribution from $m/z$ 207 is still present, but it is not known whether this ion is due to residual background or is really from the eluting substance. Both possibilities must be examined when interpreting the mass spectrum. The background corrected mass spectrum is sufficient for analysis by computerized search systems.

### 4.3.5. Biller-Biemann stripping technique

A useful means of improving the apparent resolution of a reconstructed gas chromatogram is by use of the Biller-Biemann algorithm. It is a good illustration of the advanced use of mass spectrometer scan data and in particular mass chromatograms. The technique identifies all $m/z$ values that maximize at each scan number, and strips away all other ions from that scan. This procedure not only effectively removes background ions, but also removes ions in a mass spectrum due to closely-eluting unresolved GC peaks.

The $m/z$ abundance data given in Fig. 4.9 illustrate how the stripping process is performed. The ion at $m/z$ 57 is present in all scans from 581–595, however, it maximizes only at scan 588. Therefore, $m/z$ 57 would be stripped from all scans except scan 588. When this process is repeated for all scans from 581–595, the end result is a single peak one scan wide which contains all $m/z$ values present in the mass spectrum of a single compound. In actual practice, the concentration of all ions do not always maximize at the same scan because of changing ion source concentrations of the GC peak during each scan. Therefore, it is appropriate to also leave ions in the mass spectra of the few scans obtained just before and immediately after the scan in which abundance of the $m/z$ value is maximized.

An example of the use of this stripping technique is shown in Fig. 4.16. The GC peak appears to be much better resolved after the stripping process. Three unresolved components between scans 170 and 190 are baseline resolved after stripping, and the large peak near scan 120 is shown to be actually composed of two unresolved components. Because of the method of background subtraction using mass chromatograms from consecutive $m/z$ values across the set scanning limits of the mass spectrometer, the stripped RGC is called a mass resolved gas chromatogram.

## 4.4. Compound identification using reference spectra matching

For a specific set of experimental conditions, the mass spectrum of a molecule is like a fingerprint. Since each mass spectrum is a pattern of ion fragments originating from a compound of definite structure, it is possible to identify the compound starting from a mass spectrum. Except in simple cases, this task is complex, and other data are required to confirm any proposed structure. By comparing the mass spectrum with others in a reference file, however, the task is greatly simplified. If the unknown spectrum can be found in the reference file, then the compound must have the same or very similar structure to that of the reference compound.

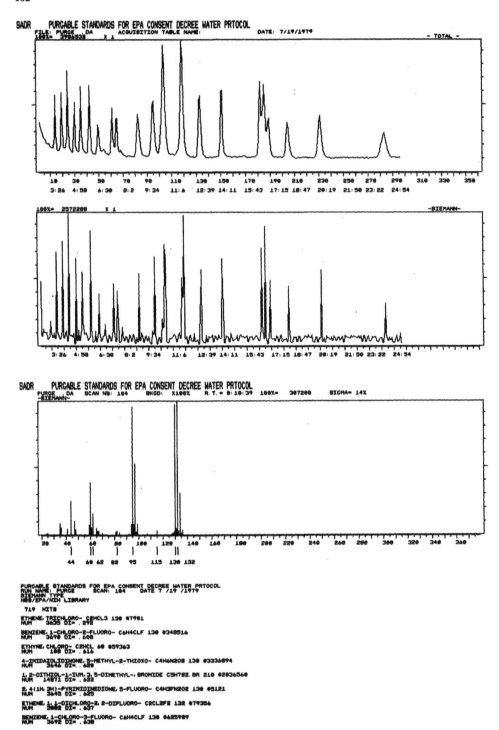

Fig. 4.16. Illustration of Biller-Biemann stripping technique showing original RGC (top) and mass resolved gas chromatogram.

Printed reference compilations are available in which the unknown mass spectrum can be matched to reference spectra listed according to the most abundant ions or by molecular weight. Manual searching is tedious, and many close matches of spectra may be missed during visual comparison. Therefore, the comparison process is usually performed using a computer.

Systems are available in which computer matching can be performed using remote computers accessed via telephone lines. Additional interpretation aids available with some computer systems are calculation of closeness of the match, identification of components of mixtures, and determination of some structural groups of the unknown even if the unknown is not represented in the reference file. The major limitation of compound identification by reference spectra matching is the available reference files. Although they are continually being updated, the largest of these files contains the mass spectra of less than 70,000 different compounds. Although this data base seems large, the number of known organic compounds is still much greater.

### 4.4.1. Mass spectral compilations

Many collections of mass spectra have been compiled, although several of these are specialized and only contain a few thousand spectra of specific compound classes such as drugs. A limitation of available compilations is that the rate of addition of new spectra is very low. Reasons for this are readily apparent; obtaining verified, accurate mass spectra free of interferences is a difficult and laborious task, and mass spectra of new compounds not already represented in available data compilations are difficult to identify unambiguously. Two published collections especially useful for interpretation of mass spectra are the eight-peak index and the Environmental Protection Agency-National Institute of Health (EPA-NIH) mass spectral data base.

*Eight-peak index.* The eight-peak index is a compilation of mass spectra condensed and organized by the eight most intense peaks in each mass spectrum, normalized to the largest as 100%. Over 31,000 mass spectra are in the eight-peak index, and they are indexed by molecular weight and elemental composition in addition to the most abundant ions. For some compounds, more than one mass spectrum has been included, therefore the total number of different compounds is less than 31,000. Because mass spectra obtained from different mass spectrometers and using different conditions may not be identical, the probability of identification is increased when multiple spectra are included in the data base.

For identification of mass spectra, a listing of peaks in order of the most abundant ions is used. Fig. 4.17 shows the data format. Once the $m/z$ of the base peak is located in the eight peak index, remaining peaks are matched in order of abundance until the identity of the unknown compound is limited to a small number of possibilities. Intensities do not have to match exactly for a good fit. The experience of each analyst is important in determining the amount of variation that can be allowed before rejecting a reference spectrum as a possible match. It is not always necessary to determine a single unambiguous structure for the unknown compound. For some applications the identification of compound class or a few

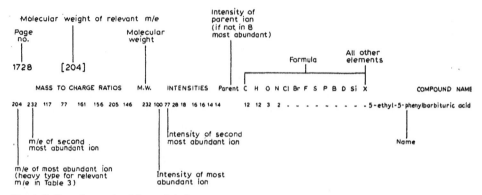

Fig. 4.17. Data format in eight-peak index.

selected structural features is sufficient. If a close match between the unknown spectrum and a spectrum in the eight-peak index is found, positive identification still requires comparison of the full spectra, since the eight largest peaks are not always the most structurally significant. For isomeric compounds having similar mass spectra, additional information such as retention time or retention index is needed. Because the parent ion is so important in the identification process, it is listed in the eight-peak index even if it is not one of the eight most abundant.

*EPA-NIH mass spectral data base.* The EPA-NIH mass spectral data base is a compilation of mass spectra of 25,556 different compounds, with no duplication. Full spectra are plotted in bar-graph format and are presented in order of increasing molecular weight. Fig. 4.18 shows the data format of these mass spectra. Since these indexes are used primarily to find the mass spectrum of a known substance, they are of limited value for the identification of mass spectra unless the molecular formula or molecular weight are known. The EPA-NIH collection, however, is complementary to the eight-peak index. Combined use of the eight-peak index and EPA-NIH printed compilations can provide effective identifications of unknowns, if mass spectra of these compounds are included in the reference collections. The EPA-NIH collection is available in computer format for use in computerized search systems.

An index for the EPA-NIH compilation lists spectra by substance name, molecular formula, molecular weight, and Chemical Abstracts Registry number. The original volumes were published in 1978. A supplement was issued in 1980 and contains an additional 8807 mass spectra. This rate of growth of about 4500 new, verified mass spectra per year seems large, but actually represents a small fraction of the total number of organic compounds not yet included in the data base. The usefulness of printed data collections of mass spectra for manual comparison is essentially limited to the specific compounds represented in these compilations. Furthermore, manual spectra matching is slow, laborious, and error prone.

*Registry of Mass Spectral Data.* Published in 1973, this collection contains the mass spectra of 18,806 different compounds. The Registry was designed to find mass spectral data for specific compounds, rather than the opposite problem of identifying substances from their mass spectral data.

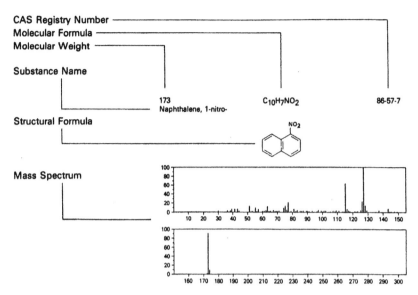

Fig. 4.18. Data format in EPA-NIH mass spectra data.

Fig. 4.19 shows the data format. Spectra are ordered by nominal molecular weight. Compounds with the same molecular weight are sorted according to elemental composition in decreasing order of carbon and hydrogen atoms, followed by increasing numbers of other elements arranged alphabetically. An index is given which references mass spectra according to elemental composition.

The Preface to the Registry illustrates the difficulties involved in developing such reference collections by describing the extensive quality control measures which were taken to ensure only the best available mass spectra were included. These

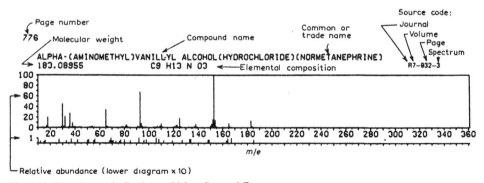

Fig. 4.19. Data format in Registry of Mass Spectral Data.

measures included checking all spectra by a computer program for three common indications of poor data: impurity peaks at masses above that of the molecular ion, isotopic abundance ratios that do not correspond to those expected from the elemental composition, and the presence of illogical neutral losses. Since each mass spectrum contained an average of 74 peaks, the authors state that there is a good possibility that some errors still exist, even after extensive checking.

*Wiley-NBS mass spectral data base on magnetic tape.* This data base has been prepared specifically for the needs of computer research and retrieval systems and is available only on magnetic tape. The first edition, released in 1984, is the largest reference collection of mass spectral data available, and contains over 80,000 spectra of more than 68,000 compounds. It combines information from the EPA-NIH collection and the Registry of Mass Spectral Data. Although not designed for manual searching as are the previously described reference collections, individual spectra can be plotted for visual comparison with unknowns. This would normally be done after a computerized search has reduced possible matches to only a few. Additions to this collection will be made periodically as new mass spectra are obtained.

### 4.4.2. Methods of computerized mass spectral search

Many of the problems with manual matching of mass spectra are reduced or eliminated by computerizing the comparison process. The computer can perform rapid and accurate matching, and the basis for comparison can be complex. A useful feature is the calculation of factors which are used to distinguish between good, average, and poor matches. A human will generally stop searching at the first good match, but a computer can be programmed to find all possibilities and list as many as desired. If a close match is not found, some information may still be obtained from the search, since the compounds of closest match usually have some structural features in common.

Many different methods of computerized search have been developed. Although the final goal of identifying compounds from their mass spectral patterns is the same for all of these methods, the approaches taken differ depending on the specific problems to which each method is applied. In some applications, search speed may be the most important consideration. For others, limitations in computing power or the size of stored reference files are key considerations. Additional factors may include flexibility in expanding or modifying the supplied data files, or adapting the search to take into account other data such as retention times. Computer search methods can be generally categorized by the following classifications.

*N most intense peaks.* This is the simplest type of computer search. It is performed by matching the $N$ most abundant peaks in the unknown mass spectrum with corresponding peaks of each mass spectrum stored in a reference file. $N$ generally ranges from four to ten. To save storage space and reduce search time, reference spectra can be reduced to the required $N$ largest peaks. The eight-peak index is an example of a reference file where $N = 8$. When comparing unknown and reference mass spectra, the relative abundance of peaks is usually taken into account, since a good match requires not only the same peaks to be present, but also in the correct relative proportions. Because some variation in relative abundances occurs in mass spectra of a substance obtained on different instruments, it is necessary to allow an intensity range of each peak used for matching.

Since several of the reference spectra may closely match the unknown mass spectrum, a means of listing them in order of best match is needed. This is accomplished by calculating a match factor or similarity index for each comparison. Best matches are listed at the conclusion of the search in order of decreasing match factor. As an example of one way in which a match factor (MF) could be calculated, consider the case where $N = 10$. For each peak compared, ten points are allocated. One point is subtracted for each five percent difference between the intensity of corresponding peaks in the unknown and reference spectra. If a peak is missing from the reference spectrum, ten points are subtracted. For a perfect match, $MF = 100$, while $MF = 0$ if no peaks are in common. During the search, the MF for each reference spectrum is calculated, and if greater than a value set by the analyst, stored in a list with the reference spectrum identification. For the MF calculation described here, a value of 80 or more indicates a good match. Instead of setting a threshold value for the MF, it may be more convenient to simply save the ten best matches.

Although a largest $N$ peak search is fast, simple to program, and adaptable to less advanced computing devices, it has the disadvantage that the largest peaks are often not the most structurally significant. A good example of this is for the mass spectra of $n$-alkanes. Members of this homologous series have almost identical mass spectra under electron impact mass spectrometry. They are best distinguished by their molecular ions, which differ by mass increments of 14 for consecutive members of the series. The molecular ions, however, will not be included in the ten largest peaks, and may not be observed for alkanes of higher molecular weight. More complex methods of computerized search have been developed to give higher weighting to the more significant peaks.

*N most significant peaks.* One method of choosing peaks of higher significance for a search is to give greater emphasis to peaks of higher mass. This is valid because as the mass of an ion fragment increases, there are fewer compounds that could have given rise to such a fragment. By defining significance as (mass × abundance), a search using the $N$ most significant peaks will be more diagnostic than one that chooses the $N$ largest peaks. Other characteristics of this search method, such as calculation of match factor, are the same as described for the $N$ largest peak search.

*N largest peaks in each 14 a.m.u.* Unlike the first two methods, the number of peaks chosen for searching using this method is variable. With $N = 2$ and if the 14 a.m.u. divisions begin at a.m.u. = 6, the search is called a Biemann search, named after the scientist who developed this technique. By choosing two peaks in each 14 a.m.u. division, the average number of peaks used for the search will be approximately 25% of the number contained in the original mass spectrum. Since peaks are chosen throughout the mass spectrum, the most structurally significant peaks will be included in the search. The 14 a.m.u. division is chosen because the -$CH_2$- unit is a basic building block of organic aliphatic molecules. An important feature of the Biemann search is that the peaks chosen must include the molecular ion, if present, because the heaviest fragment ion cannot be in the same group as the molecular ion except for losses due to $H^+$.

*Whole spectrum search.* A simple means of ensuring that all significant peaks are used for the computer search is to use the entire mass spectrum. Search times will be longer and reference files cannot be abbreviated to save storage space. Also, the basis of comparing unknown and reference spectra, and calculation of match factors, tend to be more complex with full spectrum search methods. Nevertheless, some full spectrum search procedures are capable of determining some information about the unknown, even if it is not represented in the reference file. All of the peaks in the unknown mass spectrum do not have to be in the reference spectrum for a close match, since some low abundance peaks may be from general background. The principal advantage of full spectrum search is that all important structural peaks are retained. More accurate matching can be performed, which is important in cases where several reference spectra are similar to the unknown.

*Reverse search and residual spectrum.* The reverse-search method was developed for use in full-spectrum comparisons. Although the basic concept of reverse search is simple, it has been found to increase the reliability of identification when unknown mass spectra contain a significant proportion of background peaks, or when the mass spectra of two or more compounds are mixed together because of insufficient chromatographic separation.

The basis of reverse search is that for each comparison with a reference mass spectrum, the peaks in the reference are matched with corresponding peaks in the unknown. In the ideal situation where reference and unknown spectra match perfectly, it makes no difference which way the comparison is performed. Real unknowns, however, often have impure mass spectra from background ions or closely eluting compounds not completely separated by the chromatographic column. For these situations, comparing the unknown to the reference spectrum, even if the reference is of the same compound, will not give a good match, since the unknown will contain many extraneous peaks not found in the idealized reference spectrum. Reverse searching, however, only compares peaks in the unknown that are found in the reference spectrum. Contaminant peaks from background or co-eluting compounds will be ignored.

Once a good match has been made using a reverse search, the reference spectrum can be subtracted from the original unknown mass spectrum to give a residual spectrum. This residual spectrum can be treated as a new unknown and processed through the search system for identification. Therefore, by using the reverse search method it is possible to identify more than one component of a co-eluting peak. Because of many factors relating to background peaks, reproducibility of mass spectra, and signal-to-noise considerations, it is seldom possible to identify more than two overlapping components.

*Combined match factor–retention time identification.* When identifying compounds using GC–MS data, retention times are often overlooked. Some computer search systems, however, have combined retention time or retention index values with mass spectral match factors to give an overall index for identification. Such systems are especially valuable for the identification of isomeric substances that have almost identical mass spectra but different retention times. Because of the numerous types of GC columns available, both packed and WCOT, search systems that use retention data are often custom made for individual laboratories for specific analytical needs of well-defined procedures having constant operating conditions.

*Statistical-based comparison.* The most complex computer search systems require much greater processing power than do the other search methods described above. In some cases, this means that large time-sharing computers are required instead of the computer supplied with the GC–MS instrument. Reliability of identification for any of the search methods will be good in the ideal situation where the unknown spectrum matches exactly with a mass spectrum in the reference file. The more complex statistical-based systems, however, are superior when ideal conditions are not achieved.

Statistical-based computer search methods are founded upon the assumption that specific mass fragments and patterns of these fragments depend upon specific structural features of molecules. These structural features can be expressed in terms of probability. That is, a specific pattern observed in an unknown mass spectrum can be related to the probability of presence of a specific structural feature in the unknown. For example, a mass loss of 15 a.m.u. from the molecular ion is almost certainly indicative of the presence of the structural unit $-CH_3$ in the mass spectrum.

Computerized search systems that use such statistical approaches can provide valuable information even if the unknown compound is not represented in the reference file. Although the identity of an unknown may not be determined, its compound class or other specific structural features can be inferred.

*Summary of search methods.* The principal features of the various computerized methods of mass spectral search are listed in Table 4.5. Even the simplest of these methods can work effectively if a pure, undistorted mass spectrum of the unknown is obtained, and if the unknown is found in the reference collection that is searched. Unfortunately, the rate of growth of available reference files is very slow. Therefore, more complex methods that can give some information about the unknown are often needed, even though positive identification of the unknown is not achieved. Sometimes computer searching is used even when analyzing mixtures of known compounds to ensure correct quantitation of peaks is performed during unattended automatic analysis.

Basic search strategies have reached a sophisticated level with the statistical-based procedures that have been described. In addition to expanding available data files, future systems will incorporate other information into the search such as spectra of the unknown obtained by other spectroscopic techniques. An ideal combination still in the early stages of development is GC–MS–Fourier transform infrared spectroscopy. No system, however, can give a completely unambiguous identification of an unknown based on search data only. The importance of human interpretation of mass spectral data has not been reduced. An understanding of different search strategies is necessary to evaluate the quality of data obtained from the various commercially available systems, and detailed understanding of mass spectra and how they are formed will maximize the benefits of computer search data.

TABLE 4.5

COMPARISON OF COMPUTER SEARCH METHODS

| Method of search | Basis of comparison | Reliability of identification | Other features |
|---|---|---|---|
| N most intense peaks | Match largest peaks in unknown with corresponding peaks in each reference spectrum. | Fair if unknown has limited fragmentation and is in the reference file | Fast, easy to program, good for systems with limited computing facilities |
| N most significant peaks | Same as above but largest peaks after peak mass $\times$ peak abundance | Better than above but still limited to compounds in the reference file | Same as above |
| Biemann search | Divide spectrum into 14 a.m.u. divisions, take 2 largest peaks each division beginning at a.m.u. = 6 | Excellent if unknown is in reference file. Compounds of closest match may have similar structure to unknown | About 25% of peaks used for search. Molecular ion must be chosen if present |
| Whole spectrum search | Use every peak in unknown for search | Excellent if unknown is in reference file. Poor if unknown spectrum contains contaminant peaks | Slow, efficient programming required and large data storage for reference files |
| Reverse search | Whole spectrum search but treat reference spectra as unknowns. Each comparison based on peaks in the reference spectrum only | Better than whole spectrum search for impure spectra | Identification of multiple components of mixtures can be achieved by using residual spectrum |
| Combined match factor–retention time/index | Retention time or retention index of unknown used in combination with a search procedure | Good, can be used to differentiate between compounds having similar mass spectra | System must be setup for specific chromatographic conditions |
| Statistical methods | Probabilities of specific mass fragments or patterns | Excellent, depends on detailed knowledge of mass spectral characteristics | Some information obtained even if unknown is not in reference file. Programming required is complex |

### 4.4.3. Commercial mass spectral computer search systems

Many computerized search systems have been developed, but only a few are generally available. These are based on the principles described in section 4.4.2, and are either supplied by manufacturers of GC–MS instruments or are used by transmitting data to large time-sharing computers over telephone lines.

*In-house dedicated search systems.* Manufacturers of GC–MS–computer systems generally supply a mass spectral search program with the computer software. The search strategies used vary greatly depending upon the supplier and the specific computer and computer storage devices available. For configurations consisting of smaller computers or even advanced calculators where data storage is on tape or floppy disk, data storage space and search time are primary considerations. Libraries of reference spectra on such systems are limited to the $N$ largest or $N$ most significant peaks, and the search program is appropriately designed. Complex search strategies need complex and lengthy programming that require more advanced computers to function effectively. For specific applications on limited systems, libraries are sometimes reduced to specific classes of compounds, such as the U.S. EPA list of priority pollutants.

Data systems based on minicomputers with hard-disk storage capacity of five to ten megabytes or more can support sophisticated search systems using full-spectrum search and libraries consisting of tens of thousands of reference spectra. The EPA-NIH mass spectral data base described previously is an example of the type of library available with such systems.

An important advantage of in-house systems is the possibility of expanding the supplied reference file to include new compounds of specific interest to each individual user. In some cases it is even desirable to include mass spectra of compounds that have not yet been identified, so that a record can be kept of the type of sample in which the unknown species were detected. Provision for this can be made by the users of in-house data systems.

*Mass spectral search system (MSSS)*. MSSS is an interactive system which is available through a telephone hookup with a remote computer. Only a special password obtained from the operators of MSSS, a telephone, and a computer terminal interfaced to a modem are needed. The modem is a device which permits bi-directional communication between computers over the telephone lines. MSSS is one component of the Chemical Information System (CIS) which was developed as a joint project between the U.S. NIH and the U.S. EPA. A yearly subscription fee is charged to users of CIS, plus an additional charge for computer time used.

Options for mass spectral search include the Biemann search (KB) and a peak-by-peak search (PEAK). For a PEAK search, mass peaks are entered one at a time, and an intensity range for each peak is selected by the user. After entering each peak, the compounds matching that peak (and peaks already entered) are selected by the search program. As more peaks are entered, the number of matching mass spectra are reduced. When the matches have been reduced to a few, generally five to ten, they can be listed by the user according to the closest match. This procedure is similar to an $N$ largest peak match, except that any peak can be chosen, and information is supplied after each peak entered. Some skill is required in entering peaks. Correct choice of peaks to use will not only improve the probability of obtaining a close match, but will minimize the number of entries needed to reduce the number of possible matches to a manageable number. PEAK is an interactive search because the user can choose to enter another peak, delete a peak, print results or quit the search after each peak entered.

For the KB search, all of the peaks are entered at the start, and no additional user interaction is required until the completion of the search. Very careful data entry is required, since only the two largest peaks in each 14 a.m.u. division are entered. It is advisable to list all peaks of the unknown mass spectrum on a special

TABLE 4.6
NUMBER OF OCCURRENCES OF VARIOUS MASSES IN MSSS REFERENCE FILE OBTAINED
BY USING MSSS-PEAK OPTION

| $m/z$ with abundance 70–100% | Number of mass spectra in data file |
|---|---|
| 57 | 1593 |
| 91 | 1147 |
| 149 | 224 |
| 202 | 95 |
| 252 | 62 |
| 322 | 31 |

form that is already divided into the required 14 a.m.u. sections before calling the remote computer. Since more of the mass spectrum is used, more reliable results can usually be obtained from KB than from PEAK. An experienced user, however, who is skilled at mass spectral interpretation and knows which peaks to choose, may prefer the interactive approach of PEAK. A similarity index is calculated for a KB search, from 0–10 for a perfect match, and printed results are listed in order of decreasing similarity index.

*Probability based matching (PBM).* PBM is the most successful available search system yet developed for mass spectral identification. It is available over telephone lines from Cornell University, where it was developed, and at least one commercial manufacturer supplies PBM with its GC–MS software. The major features of PBM are the use of reverse-searching and a statistical-based method of weighting mass and abundance values.

To illustrate the importance of mass values for diagnostic purposes, Table 4.6 was generated using MSSS. Since the number of compounds in the MSSS data base which contain each peak entered is reported after each entry, the total number of mass spectra containing any $m/z$ value and within any abundance range can be easily determined. Table 4.6 is a list of the number of spectra which contain the listed $m/z$ values with a relative intensity of 70–100%. It is observed that as the $m/z$ value of each peak increases, the number of mass spectra in the data base containing that peak decreases. In fact, the probability of a specific $m/z$ value decreases by a factor of two about every 130 a.m.u. This does not mean that every $m/z$ value is less probable than all other lower $m/z$ values, since some mass combinations are unlikely due to the finite possibilities in which atoms can be combined.

The probability that a specific peak will appear in a mass spectrum also depends upon its abundance. For example, 4544 mass spectra were found by MSSS which have a peak of $m/z$ 71 with an abundance of 0–100%. Only 466 spectra, however, contained this peak with an abundance of 70–100%. PBM weights peaks in a mass spectrum according to both mass and abundance values for matching purposes. PBM is very successful at identifying unknowns if their spectra are in the reference file. If a definite identification is not made some information can still be obtained, since the compounds of closest match often have structural similarities to the unknown. Because an option is available to subtract the spectrum of closest match from the unknown to obtain a residual spectrum, PBM can sometimes identify more than one component of a multi-component peak.

*Self-training interpretive and retrieval system (STIRS).* The computer search systems discussed previously (PBM, PEAK, KB) can be classified as retrieval systems, because they identify unknown mass spectra by retrieving a spectrum of identical pattern from a reference file of known compounds. STIRS was developed to deduce structural information about unknowns when the retrieval process is unsatisfactory. It is an interpretative program designed to complement PBM and is also available over telephone lines from Cornell University. Where PBM retrieval is not sufficient for identification of unknowns, STIRS can provide information concerning the presence of specific common structural features. Since STIRS was designed to provide positive information, the absence of a structural feature cannot be inferred if STIRS does not predict its presence.

STIRS uses a series of match factors to retrieve compounds having common mass spectral features with the unknown mass spectrum. For each match factor, if the retrieved compounds contain similar structural groups, then these substructures are probably also present in the unknown structure. The fourteen match factors used by STIRS each represent a specific mass spectral feature such as low mass characteristic ions or large primary neutral losses. The system is self-training because specific structural features are determined on the basis of compounds retrieved by the various match factors. As an illustration of how STIRS identifies substructures, consider the match factor for large primary neutral losses. If STIRS retrieves fifteen compounds having the same large primary neutral loss as the unknown, and thirteen of these compounds contain a methoxy group substructure, then it is probable that the unknown structure also contains a methoxy group. This possibility will be strengthened if the methoxy substructure is also found by several of the other match factors. STIRS can test for the presence of 600 common substructures, and also predicts the molecular weight, rings-plus-double-bonds value, and number of chlorine and bromine atoms in the unknown.

Some incorrect substructures will generally be found by STIRS in addition to several correct ones. Therefore, STIRS results are best used to supplement other available information, not replace it. Successful interpretation of mass spectra not identified by retrieval systems requires considerable skill of the analyst, although data from STIRS can be an effective aid in this process.

*Illustration of commercial search systems.* As an illustration of the use of commercial search systems, the mass spectrum shown in Fig. 4.20 was analyzed by MSSS (PEAK option), PBM, and STIRS from a remote terminal over telephone lines. It was known that the unknown is a mass spectrum of a trimethylsilyl (TMS) derivative. The base peak at $m/z$ 73 $[-Si(CH_3)_3]$ supports this, and the large peak at $m/z$ 147 usually indicates a di-TMS compound. The molecular ion could be at $m/z$ 288, the highest mass observed in the mass spectrum. Many neutral losses of 15 a.m.u. indicates the $-CH_3$ substructure is prominent. Other mass losses of 13 and 16 or 18 a.m.u. suggest the presence of $-CH-$ and oxygen. An isotope cluster at $m/z$

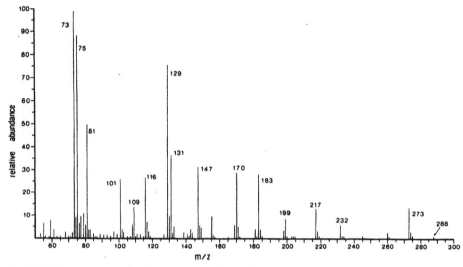

Fig. 4.20. Mass spectrum of unknown compound.

273, 274, and 275 indicates that the unknown structure contains two silicon atoms. After correcting for the contribution of these to the $m/z$ 274 peak, the remaining abundance shows that approximately ten carbon atoms are present in the $m/z$ 273 peak.

Only three peaks were entered using MSSS-PEAK option to reduce the number of possible matches to two. The peaks (and intensity ranges) entered were $m/z$ 273 (5–20%), $m/z$ 217 (8–18%) and $m/z$ 183 (5–50%). Of the two compounds retrieved, one with molecular formula $C_{10}H_6Cl_6$ could be immediately rejected since no chlorine isotope cluster was observed in the unknown mass spectrum. The other compound retrieved was an epoxy with the formula $C_{15}H_{26}N_2O_6Si_2$ (molecular weight 386). Its mass spectrum was located in the printed EPA-NIH indexes and was considerably different than that of the unknown. Higher mass peaks were used for the MSSS search since they are generally more diagnostic, however, better results may have been obtained by extending the intensity ranges for the first two peaks entered.

Fig. 4.21 shows the results of a PBM search for the unknown. All of the peaks in the mass spectrum were used for the search. PBM prints the best five matches according to their reliability. CL4 is the probability expressed as a percentage that the unknown has a structure closely related to that of the retrieved compound. CL1 is the probability that the retrieved compound (or stereoisomer) has a structure identical to the unknown. The best match found here could not be identical to the unknown, since its molecular weight is too low. However, the closest matching structures all contain oxygen and silicon. The significant residuals in Fig. 4.21 are major peaks in the unknown spectrum not present in the best matching reference spectrum. Low abundance peaks such as those at $m/z$ 274, 275 and 288 are not included in this list. PBM data in addition to earlier information supports the

39388 SPECTRA WERE SEARCH (3447 RELEVANT) BY PBM:
20 SPECTRA WERE RETRIEVED (K >= 20)

| Serial # | Name<br>MW    Formula | Reliability<br>CL4    CL1 | Contami-<br>nation | Component |
|----------|-----------------------|---------------------------|--------------------|-----------|
| #19454 | Trans-cyclohexane-1,3-diol<br>TMS (Part Cis<br>MW: 260, $C_{12}H_{28}O_2Si_2$ | 23    9 | (69%) | (32%) |
| # 5910 | Proprionic Acid-mono TMS<br>MW: 146, $C_6H_{14}O_2Si$ | 16    5 | (64%) | (74%) |
| #35967 | Monotrimethylsilyl Propionic<br>acid<br>MW: 144, $C_6H_{14}O_2Si$ | 14    5 | (64%) | (74%) |
| #35888 | Monotrimethylsilyl acrylic<br>acid<br>MW: 144, $C_6H_{12}O_2Si$ | 14    5 | (41%) | (77%) |
| # 5642 | Acrylic Acid-mono TMS<br>MW: 144, $C_6H_{12}O_2Si$ | 14    5 | (41%) | (77%) |

Spectrum #19454 Being Subtracted

List of Significant Residuals

| M/E | ABD. | M/E | ABD. | M/E | ABD. | M/E | ABD. | M/E | ABD. |
|-----|------|-----|------|-----|------|-----|------|-----|------|
| 108 | 600  | 109 | 1400 | 129 | 6457 | 130 | 771  | 131 | 3371 |
| 133 | 600  | 181 | 400  | 183 | 2900 | 184 | 400  | 198 | 400  |
| 199 | 900  | 232 | 600  | 233 | 100  | 234 | 100  | 273 | 1400 |

Fig. 4.21. Results of PBM search of unknown mass spectrum.

presence of the –OSiTMS substructures. Since the CL4 reliability is only 23% for the closest matching compound, it is unlikely that the reference data file contains the unknown compound. The percent contamination and percent component are measures of the purity of the unknown mass spectrum. Percent contamination (69%) approximates the total ionization of the peaks in the unknown that are not contained in the reference spectrum. The percent component (32%) is at the smallest ratio (abundance of unknown/abundance of reference) of all the peaks used in matching. These numbers are approximations and do not necessarily add to 100%. A PBM search of the residual spectrum did not produce any close matches.

Fig. 4.22 shows the results of a STIRS analysis. STIRS retrieved fifteen compounds with similar mass spectral features based on fourteen match factors. The substructures $-CH_3$, $-Si(CH_3)_3$, and $-O-Si(CH_3)_3$ were identified with 99% reliability. Nine of the fifteen retrieved compounds contained these substructures. Possible presence of the $-OCH_2-$ substructure is also suggested, but with lower

116

| Reliability | Substructures description | ID# |
|---|---|---|
| 99% | $-O-Si(CH_3)_3$ | 3 |
| 99% | $-Si(CH_3)_3$ | 8 |
| 99% | Methyl | 16 |
| 87% | $-OCH_2-$ | 101 |

*** Occurrence of substructures in the retrieved compounds

Compound #        Substructures - ID#

| Compound # | 3 | 8 | 16 | 101 |
|---|---|---|---|---|
| 19454 | + | + | + | |
| 26235 | + | + | + | |
| 19455 | + | + | + | |
| 19453 | | | | |
| 29930 | | | | |
| 30698 | + | + | + | + |
| 30942 | | | | |
| 29277 | + | + | + | + |
| 26234 | + | + | + | + |
| 13699 | | | | |
| 30587 | + | + | + | |
| 29933 | | | | |
| 32130 | + | + | + | + |
| 32584 | + | + | + | + |
| 19458 | | | | |

** the most probably number of carbon atoms for the unknown, N, is
     0    N    10.
   K-value for the prediction is 8.

The most reliable elemental composition among 3 possiblities based on molecular
weight 288 is
     C 12  H 24  O 4  Si 2

| K | Elemental Composition | | | |
|---|---|---|---|---|
| 26 | C 12 | H 24 | O 4 | Si 2 |
| 8 | C 14 | H 25 | O 4 | P 1 |
| 3 | C 16 | H 32 | O 4 | |

Combination 1,2,3,4,7 Match : MF11.3

| MF 1 | MF 2A | MF 2B | MF 3A | MF 3B | MF 4A | MF 4B | MF 5A | MF 5B | MF 6A | MF 6B | MF 5C | MF 6C | MF 7 | Overall MF 11 .1 | .2 | .3 | .0 |
|---|---|---|---|---|---|---|---|---|---|---|---|---|---|---|---|---|---|
| 698 | 602 | 665 | 773 | 551 | 444 | 472 | 605 | 166 | 390 | 196 | 0 | 0 | 0 | 610 | 339 | 458 | 489 |

#19454   Trans-cyclohexane-1,3-diol TMS (part CIS)
L6TJ AO-Si-1&1&1 CO-Si-1&1&1 -T
MW 260 $C_{12}H_{28}O_2Si_{12}$

Fig. 4.22. Results of STIRS search of unknown mass spectrum.

CORRECT
STRUCTURE

COMPUTER
CLOSEST MATCH

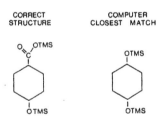

Fig. 4.23. Identified structures of unknown compound and closest PBM–STIRS match.

reliability (87%). Based upon a molecular weight prediction by STIRS of 288, the most reliable elemental composition was determined to be $C_{12}H_{24}O_4Si_2$. The names and formulas of fifteen retrieved compounds with their match factors for each of fourteen characteristic mass spectral features were printed, but only the top three are shown in Fig. 4.22. STIRS chose the same compound as PBM for closest match. It is not expected that the compound retrieved by STIRS with the highest match factor is identical to the unknown, since PBM would have found it in the reference file with a high reliability of identification.

The unknown mass spectrum shown in Fig. 4.20 was a true unknown, and the interpretation steps which have been described, including use of commercial search systems, are typical of those needed to aid identification. Although this unknown was not in the reference file, the commercial search systems were able to confirm the compound as a TMS derivative. The closest matching compounds were cyclohexane ring structures. A literature search later revealed the true identity of the unknown to be very similar to the PBM-STIRS closest match. These structures are shown in Fig. 4.23. Although the ability of computer search systems to supply useful information has been demonstrated, considerable skill is still required on the part of the user to obtain the maximum benefit of the supplied data in cases where the unknown is not found in the reference file. The analysis of unknown mass spectra should include manual inspection and any other chemical or spectroscopic data available.

*Future of computerized interpretation.* Other computer interpretation methods have been developed. Many of these are based on pattern recognition techniques, and have shown limited success so far. As more is known concerning mass spectrometer fragmentation rules, such approaches may eventually lead to the development of complete mass spectral interpretation by computer, however, the problem is so complex that this is unlikely to occur for some time. Continuing expansion of available reference files will increase the applicability of search systems such as MSSS and PBM-STIRS.

## 4.5. Quantitative analysis by selected ion monitoring

GC–MS is unique in its ability to quantify trace levels of specific organic compounds in the presence of hundreds of other substances, some of which are present in concentrations thousands of times higher than the analyte. This capability can be achieved because of the unique and complementary nature of GC and MS:

GC isolates the analyte from most of the other components in a mixture and MS provides selective and sensitive detection. To achieve low detection limits, GC peaks must be narrow (high column efficiency) and a characteristic ion (or ions) of the analyte monitored using SIM.

Quantitation can be performed using peak heights or areas from reconstructed ion plots. These ion plots can be made from data obtained with the mass spectrometer operated in either scanning or SIM mode, however, the superior detection limits achievable with SIM makes it the technique of choice for any application in which quantitation of known compounds is performed.

Because of many recent applications that require quantitation of specific compounds present at trace levels in complex mixtures, the GC–MS–SIM technique has grown in importance. Many uses involve the measurement of toxic substances in the environment. Other applications in the forensic sciences, chemical research and study of disease are equally important. Regardless of the compound or type of sample analyzed, an understanding of how SIM data are obtained and used is needed to obtain the maximum benefit from this powerful technique.

### 4.5.1. Choice of ions: basic considerations

A description of SIM has been presented in section 4.3.4. Since only a few selected ions are used in a SIM analysis, and because these are chosen before a sample is injected, careful consideration must be given to which ions are monitored. The principal requirements for quantitation are that the ions chosen are specific to the compounds quantified and they should be principal ions in the mass spectra of these analytes. More than one ion can be monitored for each compound of interest, however, detection limits become increasingly poor as the number of ions is increased. Usually, the base ion of a compound will be chosen. If the mass spectrum of the analyte has several ions of high relative abundance then the high abundance ion of highest mass will generally result in reduced background, and therefore a higher signal-to-noise ratio for the determination. Choosing the ions that give the greatest signal-to-noise ratio in the sample will result in achieving the best detection limits for a specific analysis. Initially, ions can be chosen from reference mass spectra of the compounds to be monitored. If interferences are encountered at these ions when analyzing real samples, different ions can be selected. It should never be assumed that real samples contain no interferences. For this reason, it is generally advised to monitor two or even three ions characteristic of a specific compound. If the analyte is present in the sample and there are no interfering compounds, then a response will be obtained for all characteristic ions monitored. Further, all of these ions will maximize at the same retention time, have the same peak shape, and will be present in known relative abundances. The possibility of interferences increases as attempts are made to achieve lower detection limits.

### 4.5.2. Magnetic sector versus quadrupole analyzers

In terms of ease of use and flexibility of analysis, the quadrupole analyzer is preferred for GC–MS–SIM applications. Because ions are filtered depending upon dc and ac voltages applied to the quadrupole rods, it is simple to rapidly switch

TABLE 4.7
POSSIBLE INTERFERENCES FOR 2,3,7,8-TCDD

| Compound | Formula | Mass of interfering ion | Resolution needed for separation |
|----------|---------|-------------------------|----------------------------------|
| Heptachlorobiphenyl | $C_{12}H_3Cl_7$ | 321.8678 | 13,000 |
| Nonachlorobiphenyl | $C_{12}HCl_9$ | 321.8491 | 7300 |
| Tetrachloromethoxybiphenyl | $C_{13}H_8OCl_4$ | 321.9299 | 8900 |
| DDT | $C_{14}H_9Cl_5$ | 321.9292 | 9100 |
| DDE | $C_{14}H_8Cl_4$ | 321.9292 | 9100 |
| Hydroxytetrachlorodibenzofuran | $C_{12}H_4O_2Cl_4$ | 321.8936 | Cannot be resolved |

between ions of widely varying $m/z$. A series of ions may be sequentially monitored with little regard to order of $m/z$ values and with each ion having a different dwell time. Quadrupoles have a linear mass scale, and are not subject to the hysteresis effects that limit some magnetic sector instruments.

The only disadvantage to using quadrupole analyzers is their low resolving power. High resolution magnetic sector or double focussing (HRMS) instruments are needed in applications where interferences may be a problem. For example, Table 4.7 shows some of the possible interferences for the quadrupole mass spectrometer (LRMS) analysis of an environmentally important compound, 2,3,7,8-tetrachlorodibenzo-p-dioxin (2,3,7,8-TCDD). Using LRMS, it would not be possible to distinguish 2,3,7,8-TCDD from the other compounds of Table 4.7 if the principal ion of $m/z$ 322 was monitored. Some of the interferences could be resolved by chromatography (i.e. different retention times). However, if any of these compounds co-elutes with 2,3,7,8-TCDD, it is possible to obtain a false positive-identification of 2,3,7,8-TCDD when it is not really present in the sample. If several ions were monitored, say both $m/z$ 320 and 322, then a co-eluting interference will make the ion ratios incorrect and a false negative may result, falsely concluding that 2,3,7,8-TCDD is not present in the sample. Using HRMS, at a resolution of 13,000, 2,3,7,8-TCDD can be distinguished from most of the compounds listed in Table 4.7, even if they co-elute.

The one exception is hydroxytetrachlorodibenzofuran, which has the same molecular formula as 2,3,7,8-TCDD, and therefore the same molecular mass. In this situation, the best means of eliminating interference is by chromatographic separation.

Multiple-ion HRMS–SIM is more difficult to perform than LRMS–SIM. The basic equation describing mass selection in a magnetic analyzer was discussed in section 3.2.2, and is as follows:

$$m/z = H^2R^2/2V$$

where $R$ is the sector radius, $H$ is the magnetic field strength, and $V$ is the accelerating voltage. SIM may be performed by changing either $H$ or $V$, or a combination of both. Unfortunately, magnetic field strength is difficult to change rapidly and reproducibly, and magnets tend to suffer from hysteresis effects. Large improvements have recently been made by the introduction of laminated magnets, but magnetic field switching cannot be performed with the same flexibility as electric field switching on quadrupole analyzers. Difficulties also exist with varying the accelerating voltage. Since $V$ varies inversely with $m/z$, smaller (and less effective) voltages are required for higher mass ions. A compromise is to vary $H$ for large jumps in $m/z$ values, and then change $V$ for small jumps and to "fine-tune" the analyzer. High resolution conditions are maintained during ion switching by comparing sample peaks with the exact masses of peaks from reference compounds that are continually bled into the ion source, usually through a direct probe inlet. Due to the increased difficulty and expense in using HRMS for multiple-ion SIM analysis, it is generally used for difficult samples that can not be easily analyzed by LRMS, and as a quality control check on LRMS results. A quality control program, for example, might include the analysis of 10% of LRMS samples by HRMS. Quadrupole systems, however, are well suited for the high sample throughput required to perform routine analysis.

### 4.5.3. Identification and quantitation procedures

To perform quantitative analysis by GC–MS–SIM, it is first necessary to ensure that the desired compounds are really those that have been detected. Although judicious choice of ions will decrease the probability of interferences, the analyst must always be alert to this possibility. In SIM analysis, only a few ions are monitored, and many applications require analysis of minor components in mixtures of hundreds of compounds. Each of these substances, some of which may be at concentrations thousands of times greater than the analytes, form many ions in the MS source. How SIM can be used to ensure compound identification is shown by the following example.

Fig. 4.24 is the total ion plot of a GC–MS–SIM analysis of compounds extracted from municipal refuse. Several specific compounds were of interest, therefore twelve $m/z$ values were used. These are listed in Table 4.8. Seven peaks were detected, as labelled on Fig. 4.24. Two of the compounds to be quantified were isomers of formula $C_{12}HO_2Cl_7$ (molecular weight 422). Three of the principal ions in the mass spectra of these isomers were previously determined from analysis of pure standards to be at $m/z$ 422, 424, and 426 in the relative abundances 44:100:97, respectively. These three $m/z$ values were among the twelve chosen for the SIM analysis.

Fig. 4.25 is a plot of the ion abundances of $m/z$ 422, 424 and 426 between retention times 6.0 min and 7.5 min of Fig. 4.24. Only three peaks are observed. Peaks A, C, E of Fig. 4.24 do not respond at any of the three masses plotted, and peak F shows no response at $m/z$ 426, therefore they cannot be the substances of interest. All three masses are represented in peaks B and D. By comparing peak areas shown in Fig. 4.25, the relative abundances of these peaks are 36:100:99 (B) and 38:100:100 (D) for $m/z$ 422, 424, and 426, respectively. Since these relative abundances or ion ratios are well within experimental deviation from the theoretical ratios, peaks B and D are identified as the two isomers of formula $C_{12}HO_2Cl_7$.

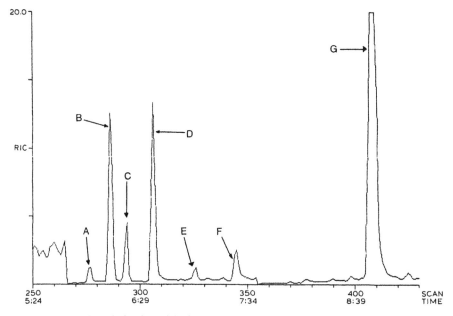

Fig. 4.24. GC–MS analysis of municipal refuse extract: total ion plot of 12 ions monitored.

Depending upon the MS calibration, relative abundances of ions and compound concentration, the measured relative abundances can vary from 10–20% of theoretical values. Frequent analysis of standards is important to measure the precision with which ion ratios can be determined. As a final step in the identification, the retention times of peaks B and D were found to exactly match those of standards

TABLE 4.8

IONS MONITORED DURING GC–MS–SIM ANALYSIS OF MUNICIPAL REFUSE EXTRACT

| Compound monitored | Formula | $m/z$ | Theoretical ratio |
|---|---|---|---|
| Heptachlorodibenzo-p-dioxin | $C_{12}HO_2Cl_7$ | 421.8 | 44 |
| | | 423.8 | 100 |
| | | 425.8 | 97 |
| Heptachlorodibenzofuran | $C_{12}HOCl_7$ | 405.8 | 44 |
| | | 407.8 | 100 |
| | | 409.8 | 97 |
| Octachlorodibenzo-$p$-dioxin | $C_{12}O_2Cl_8$ | 457.7 | 88 |
| | | 459.7 | 100 |
| Octachlorodibenzofuran | $C_{12}OCl_8$ | 441.7 | 88 |
| | | 443.7 | 100 |
| Octachloro[$^{13}C$]dibenzo-$p$-dioxin | $^{13}C_{12}O_2Cl_8$ | 469.7 | 88 |
| | | 471.7 | 100 |

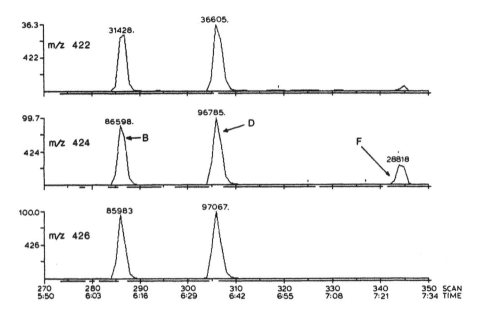

Fig. 4.25. Plot of three characteristics ions of heptachlorodibenzo-*p*-dioxin monitored during GC–MS-SIM analysis of municipal refuse extract.

analyzed using the same chromatographic conditions. Table 4.9 is a summary of the criteria used in the above example to identify compounds by GC–MS–SIM. In many applications, two characteristic ions are sufficient. Monitoring a single ion per compound is acceptable for simple mixtures, or well studied systems where no interferences have been previously demonstrated. It is better to assume interferences are present until shown otherwise.

To quantify the two compounds identified above, a standard of one of them was analyzed using the same GC–MS–SIM conditions. Only one $m/z$ value was used for quantitation. Generally the most abundant ion is used, which in the above example was $m/z$ 424. This ion gave an area of 6400 (arbitrary units) for an injection of 100 pg. The corresponding standard response factor is $6400/100 = 64$ area counts per picogram injected into the GC–MS system. By dividing the area of the $m/z$ 424 ion from Fig. 4.25 by this response factor, the amounts detected in peaks B and D were 1.4 ng and 1.5 ng, respectively.

TABLE 4.9
CRITERIA FOR SIM IDENTIFICATION

1. Exact correspondence of three characteristic ions
2. Relative abundances of ions are correct
3. Retention times are the same as standards
4. Sample was processed through selective procedure designed to remove interferences
5. Peaks were at least three times larger than background noise

For the greatest accuracy, standards should be analyzed which bracket the unknown concentration. In quantitative GC–MS–SIM work, however, multipoint calibration curves are often not determined. This is due to sample throughput considerations, since a single analysis of a complex mixture may require 1.0–1.5 h. To perform a three- or four-point calibration would require 50% or more of the total analysis time in a normal working day. It then becomes important to group sample analysis by expected concentration, either low, medium or high. For most routine work, this does not present a problem. An appropriate standard is then analyzed with each group. To improve accuracy, internal standards are employed in addition to external standards. The most effective internal standards are isotopically labelled analogues of the analytes.

### 4.5.4. Use of isotopically labelled standards

Internal standard methods of quantitation are common for GC–MS–SIM analysis. Several advantages of this technique compared to external standard quantitation are: (1) compensation can be made for variations in injection technique, (2) effects of other injection variables such as leak in the septum can be detected, (3) shifts in retention times can be determined, (4) by spiking the sample with the internal standard before extraction and other chemical treatment, the efficiency of these operations can be monitored, and correction for sample losses can be performed.

The most effective internal standards are isotopically labelled analogues of the substances being analyzed. Such compounds have the same chemical and physical properties as the corresponding analytes, and are effective monitors of sample extraction and chemical cleanup efficiencies. In addition, an isotopically labelled analogue of the analyte can be used to confirm identification by correspondence of retention times. It should be noted, however, that retention times may not match exactly, since modern high performance WCOT columns with thousands of theoretical plates can sometimes separate a compound from its labelled analogue.

Fig. 4.26 is a plot of three ions from the SIM analysis shown in Fig. 4.24. The plot covers retention times of 8–9 min, which only includes peak G from Fig. 4.24. This plot not only illustrates the use of isotopically labelled internal standards, but shows the value of SIM in analyzing substances even if they cannot be chromatographically separated. In Fig. 4.24, peak G appears to be a single large peak. It is, however, actually composed of three different compounds. The most abundant of these, corresponding to $m/z$ 460, has the same basic structure as the isomers detected in Fig. 4.25 but with the lone hydrogen replaced by another chlorine atom ($C_{12}O_2Cl_8$). Similar to this structure, the compound detected at $m/z$ 444 has one oxygen less ($C_{12}OCl_8$). The third substance detected is a $^{13}C$-labelled analogue of the most abundant compound ($C_{12}O_2Cl_8$). Since every carbon ($^{12}C$) in this molecule has been replaced by $^{13}C$, its mass is 12 units greater than that of the native compound. Ion abundances of the three compounds were easily distinguished from each other by MS, even though the compounds were not resolved chromatographically and one substance was present in an abundance which was about 100 times greater than the abundances of the other two compounds.

124

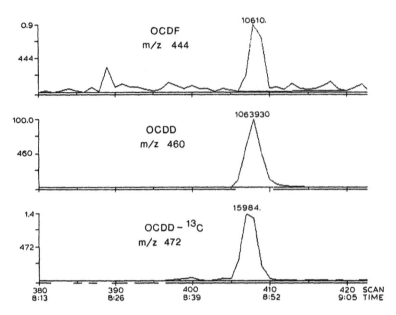

Fig. 4.26. Plot of three $m/z$ values from single GC–MS-SIM analysis showing separate detection of three co-eluting compounds including isotopically labelled internal standard.

The isotopically labelled compound can be quantified in the same manner as previously demonstrated to calculate its recovery. For example, if 1.0 ng was added to the sample before extraction, and 0.5 ng was detected in the analyzed sample, then recovery is 50% and the amount of unlabelled compound detected in the sample would be corrected by multiplying by a factor of two. It is assumed that the isotopically labelled analogue behaves in the same manner throughout the analytical procedures as its unlabelled counterpart. Another method of quantitation uses the area of the internal standard to calculate its response factor as though the recovery was 100%. The analyte is then quantified by dividing its peak area by this response factor, and multiplying by a correction factor that is the ratio of native compound and labelled compound response factors determined previously from analysis of standards. The actual recovery does not need to be calculated, since it is automatically compensated for in the calculations. It is generally assumed that corresponding native and labelled compounds have identical response factors, but this is not always the case and should be verified experimentally. In some cases response factors are quite different, such as for chlorinated compounds where the labelled analogue consists of 100% of the $^{37}Cl$ isotope.

For the identification and quantitation of picogram amounts of specific compounds, addition of an isotopically labelled analogue to the sample at some stage prior to GC–MS-SIM detection is an important analysis technique. The benefits of this are offset somewhat by the large expense of synthesizing these substances, and by the fact that only a small number are available commercially. Analysis such as demonstrated above is only possible with the GC–MS-SIM technique.

*4.5.5. Precision, accuracy and limit of detection*

Variables that affect chromatographic performance are also important in GC–MS–SIM analysis. Large amounts injected and narrow peak shapes, free of tailing are required to obtain the best detection limits. GC resolution can often be sacrificed to achieve this by employing shorter, wide-bore WCOT columns instead of narrow-bore, longer columns and by employing high GC temperature ramp rates. Increased GC resolution may be required for determinations in which interferences are present, especially for LRMS analysis. Packed column use is declining for quantitative work in spite of its greater capacity than WCOT columns because this advantage is offset by the necessity of employing a GC–MS interface such as a jet or membrane separator. WCOT columns, even wide-bore, can be installed so that the column outlet is at the ion source inlet. The direct inlet eliminates the inevitable losses that occur with separators, since all of the GC peak reaches the ion source, and also minimizes degradation of peak shape. Column bleed is not a major factor in GC–MS–SIM, unless this background contains specific ions that interfere with the analysis. The injection method used is an important factor in quantitative analysis. On-column injection is preferred for reproducibility and low detection limits. Split injection is especially poor for quantitative analysis while good results can generally be obtained in the splitless mode.

Mass spectrometer tuning is an important aspect of GC–MS–SIM analysis. Because only a few ions are monitored, it is not necessary to use conditions that are ideal to obtain mass spectral patterns that match library data file patterns. For example, most spectra in library data files were obtained with the MS ion source electron energy set at 70 eV. Using lower electron energies will generally result in more abundant molecular ions, and therefore better detection limit if molecular ions are used for SIM analysis. Another important tuning parameter is the MS peak width. This can be adjusted to be wider for SIM analysis than is usually used for full-scan MS analysis. The advantage of wider peak widths for SIM is that any minor drift in the peak position will not adversely affect the analysis. For very narrow MS peak shapes, a small shift in position will result in loss in sensitivity and reproducibility. There are no set conditions that are ideal for all GC–MS instruments, although general principles used for optimization are the same.

Operational variables include the number of ions to be monitored and the dwell time of each ion. Sensitivity will decrease as the number of ions monitored is increased and the fractional dwell time is decreased. The dwell time is the time the analyzer allows only ions of a chosen $m/z$ to reach the detector. Fractional dwell time is the proportion of total dwell spent on a specific ion.

Shorter total dwell times must be employed for fast eluting, narrow peaks such as are obtained with WCOT columns. Usually, dwell times between 0.05 and 0.5 s per ion will be chosen. Total dwell times for all ions monitored are generally 0.5–2.0 s. A good rule is to choose the number of ions and their dwell times so that at least five to ten points can be taken across each peak of interest. Another consideration is that as shorter dwell times are employed, more total data points will be obtained for an analysis, and therefore greater data storage will be required. If dwell times are too long, however, it is possible to miss fast-eluting WCOT peaks.

Other important considerations include the efficiency of the vacuum system, proper use of heated zones, and cleanliness of the ion source and analyzer. Heated zones include the injection port, GC–MS transfer line, and MS ion source. Injection port temperature depends upon the injection mode used. When employing elevated temperatures (250–300°C) care must be taken to ensure the analyte is not thermally decomposed. In special situations, it is possible for chemical reactions to occur in the injection port, and the analyst must be aware of those that could affect the analysis. The GC–MS transfer line usually is set to be at or slightly above the highest GC temperature. Any cold spots could cause condensation of the sample and must be avoided. MS ion source temperature affects the degree of fragmentation of molecules. Lower ion source temperatures are conducive to more abundant molecular ion peaks, however, the ion source will require more frequent cleaning. Maintaining a clean ion source is essential to achieving low limits of detection. The analyzer must also be clean, but this is seldom a problem if proper vacuum is maintained.

For magnetic sector instruments, the slit settings are important variables that affect the detection limits that can be attained. Narrow slits which block a large percentage of ions from entering the analyzer are needed to perform high resolution work, however, as slits are narrowed fewer ions are detected and the detection limit is increased. In HRMS–SIM analysis, there is always a trade-off between MS resolution and detection limit. In some cases, detection limits can be improved by increased resolution if the higher resolution causes a reduced level of chemical interferences. In such a situation, the signal-to-noise ratio for the analysis is improved, even though the overall signal level is decreased.

The detection limits which can be achieved also depend upon the specific application. The various compound classes have different ionization efficiencies and degrees of fragmentation and therefore different GC–MS–SIM responses. For some compounds, the best quantitative results are obtained using chemical ionization or

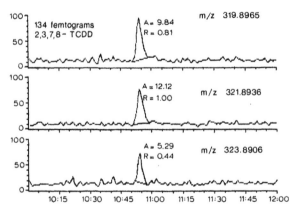

Fig. 4.27. Detection of 134 fg of 2,3,7,8-TCDD by double focussing GC–MS instrument operated at 12,000 resolution in SIM mode.

negative ion MS techniques. Quadrupole instruments can generally achieve detection limits in the low picogram range. Fig. 4.27 illustrates the detection of 134 fg of 2,3,7,8-TCDD by HRGC–HRMS–SIM (see Table 4.7). This represents about the lowest detection limits that can presently be achieved using conventional GC–MS–SIM techniques. Extending these limits to the low femtogram or better range will require complete optimization of all GC and MS parameters, and improved MS ionization efficiency.

### 4.6. Automated GC–MS operation

Automated analysis is routine in GC. Automatic injectors, commonly called autosamplers, can be adapted to almost any model chromatograph. The most advanced instruments have built-in software to control their automatic operation. Advantages of automation are obvious: technicians previously dedicated to instrument operation can perform other duties, and sample throughput can be maximized by around-the-clock instrument operation. Also, once a procedure has been set up, the level of technical expertise needed to analyze samples is lower for automated methods. With these benefits, it is perhaps surprizing that automated GC–MS analysis is not more commonplace, since the chromatograph of a GC–MS system is no different in principle than a stand-alone chromatograph, and the high cost of GC–MS systems makes sample throughput a more important consideration.

Instrument control, however, is much more complex for a GC–MS system than for a single chromatograph. Operation of an autosampler for GC–MS analysis requires advanced software and coordinated control of four devices, the autosampler, the chromatograph, the mass spectrometer analyzer, and the data storage device. Electronic interfacing between these components is available on some systems and can be developed for others. The availability of advanced software to control the system and acquire and analyze data, however, has lagged behind increased instrumental capability. Other reasons why total GC–MS automation has been slow to evolve are the following:

(1) GC–MS operation for many years has required scientist level operators and analysts for data acquisition and interpretation. This often resulted in the use of GC–MS for non-routine, difficult samples, where sample throughput was not expected to be great.

(2) Interpretation of GC–MS data using older data systems was slow and laborious.

(3) Early GC–MS work was done using packed columns and interface devices such as the jet separator. A great deal of skill and personal attention to each sample was required to obtain optimized results. Mass calibration on some early instruments was subject to drift and frequent re-calibration.

(4) Computer technology has now advanced to the stage where a dedicated computer system for GC–MS control and data analysis is both feasible and economical. Modern computers are much faster than earlier systems. Important recent advances have been made in electronic interfacing and bulk data storage devices such as disk drives.

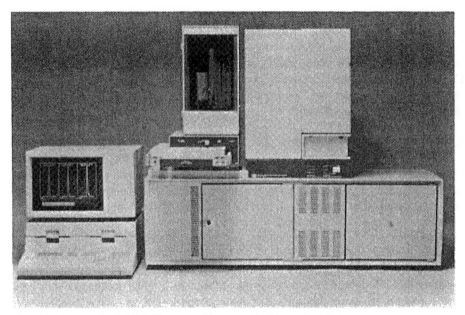

Fig. 4.28. Photograph of GC–MS system with automatic injector installed.

The rapid acceptance of lower cost quadrupole analyzers and their application to target compound analysis, especially in the field of environmental pollution, has increased the demand for greater sample throughput in GC–MS systems. Software is continually being refined to provide increased automation in data reduction. Automatic tuning, computer-controlled instrument operation and preliminary data reduction are standard features on many GC–MS systems. Various degrees of automation are possible.

### 4.6.1. Automated data acquisition

Unattended analysis of samples has three requirements: automatic injection, repetitive mass spectrometer scanning during GC operation, and rapid data transfer to a storage device. Devices employed for automatic sample injection are commonly called autosamplers. Fig. 4.28 shows one such device installed on a GC–MS system. Injection of samples is performed in much the same manner as by a human operator. A mechanical arm lifts a syringe to a vial containing the sample, pushes the syringe needle through a septum, and withdraws a pre-set amount of liquid. The sample is introduced into the GC–MS system in the same manner, by injection through a self-sealing septum. Such a system is simple, effective and reproducible, but does have some disadvantages. Techniques such as hot needle and solvent plug injection containing sample, air, and solvent layers are not used, although in principle they could be programmed. Alignment of the needle and injection port septum must be precise, or the needle could be bent when performing an injection.

The automated system will still try to analyze all samples for which it has been programmed, even if injection becomes impossible.

Another important concern is cross-contamination between samples. This can be avoided by alternately placing sample and pure solvent vials in the autosampler tray. By programming, the system will be able to recognize the difference between vials. After each sample injection, multiple syringe rinses can be performed by drawing up and discarding clean solvent. In addition, a portion of the next sample to be analyzed can be used to rinse the syringe before the sample is injected into the GC–MS.

A disadvantage with most current autosamplers is that they cannot perform on-column injections using WCOT columns. Equipment to perform this task is in the early stages of development, however, it will be difficult for automated systems to match manual on-column injection, especially for narrow-bore WCOT columns. Fig. 4.29 shows one of the first commercial WCOT automated on-column injectors. It is a mechanical injection system and operates in a similar manner as previously

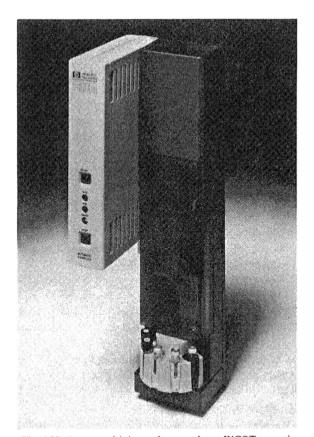

Fig. 4.29. Automated injector for on-column WCOT operation.

described for the autosampler shown in Fig. 4.28. Two principal features have been optimized to allow on-column operation; mechanical alignment of syringe and column, and very rapid injection. The autosampler has been designed with the ability to control its own parameters without adding a separate control box. Because the injector requires a metal syringe that can penetrate the septa used for sample vials, on-column injection with narrow-bore WCOT columns cannot be performed.

After injection, GC–MS control and data acquisition are no different under automated operation than for single sample, manual injection. Repetitive scans of the analyzer during the GC run are taken and stored on a bulk data storage device, usually a disk drive. For automated work, the disk capacity must be sufficient to hold all of the information generated during the period of unattended operation. Completely enclosed units called Winchester disk drives are available with data storage capacity of over 100 Mbytes.

To protect the ion source filament, electron multiplier detector and associated sensitive electronics, it is advisable to turn off the filament and electron multiplier voltage between samples and when solvent peaks elute. Diagnostic checks must be incorporated into the operating software to abort the analysis if conditions develop that could cause damage or result in invalid data.

### 4.6.2. Automated data analysis

No consideration was given in the previous section for data analysis. While this can also be automated, the software required to do an effective job of this difficult task is very complex. In principle, any problem that can be solved using pen and paper in a finite amount of time by following logical rules can be performed by a computer, however, it is difficult to define clearly such rules for GC–MS data analysis. Considerable improvements in available software have been introduced in recent years for both qualitative and quantitative automated data analysis. The key to flexibility in automated methods is to break down the data analysis problem into a number of small tasks, each of which has associated software routines. These routines can then be linked together as needed to customize data analysis for individual samples. By setting up a list of samples to be analyzed and their associated procedures, the entire process can be automated including a final analysis report. If further investigation is required, all of the raw data will still be available on disk. A typical data analysis procedure could include the following:

(1) get data file name from list - read data file from disk;

(2) find a GC peak;

(3) perform initial functions such as background subtraction, normalize spectrum, calculate ratios for SIM analysis, plot normalized spectrum and retention time;

(4) compare mass spectrum (or SIM ratios) with data base of reference spectra by library search for peak identification and include retention time (or retention index) data in identification criteria;

(5) calculate peak area for quantitative analysis;

(6) quantitate peak by comparing area with area of internal/external standard;

(7) repeat steps 2–6 until no more peaks are found;

UNKNOWN

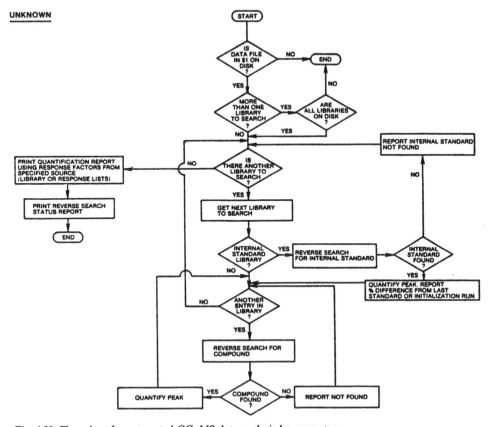

Fig. 4.30. Flow chart for automated GC–MS data analysis by computer.

(8) repeat steps 1–7 until all samples analyzed;

(9) print final report and include all sample identification and GC–MS analysis conditions in a customized header, all retention time (index) data, and compound identifications and quantitation.

A flow chart showing a program for automatic GC–MS data analysis is given in Fig. 4.30. It has incorporated the general analysis scheme described above, and is available as one of the many programs that comprise the Finnigan MAT INCOS data system (see section 4.7.1). Several data checks have been included in the analysis scheme, including checking for the accessibility of data files and libraries to be searched. Without such checks it is possible to generate an error condition preventing analysis of many data files because of a single mistake in one data file. For individual samples internal standards can be used, and the program shown in Fig. 4.30 will check for their presence. If the internal standard is not found, this indicates a problem with the sample and processing can be begun on the next data file. Library search and quantitation tasks are performed by separate programs that are called as needed by the main program.

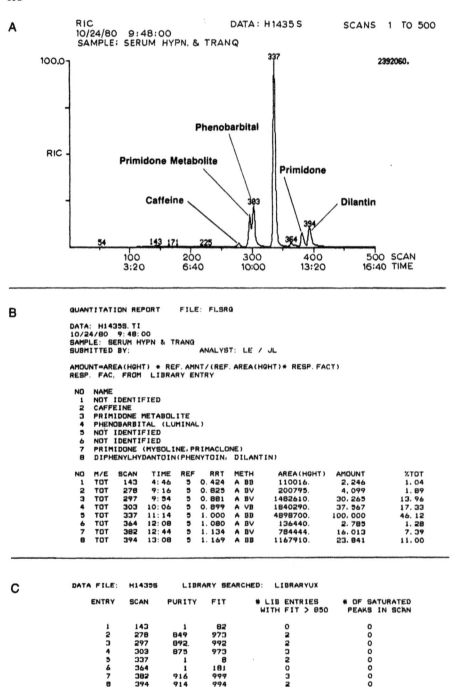

Fig. 4.31. Automated GC–MS analysis report. Compounds are identified by computer library search and automatically quantitated by comparing peak areas with areas of standards analyzed using the same conditions.

The type of data output available for such automated systems is illustrated in Fig. 4.31. This survey analysis identifies compounds using a forward library search. Confirmation of identity is based upon library search match factors. Quantitation is then performed using specific ion abundances. Various programs are used to print the background corrected spectrum and three best library matches for each GC–MS peak, integrate each peak area, and for the forward library search of each peak. The automatic report shown in Fig. 4.31 includes reliability estimates for the qualitative and quantitative data presented.

Although the data analysis illustrated above was automated, all control still resided with the user. Libraries to be searched, criteria for compound identification, and other key parameters were all set initially by the user. Automation is desired because the same steps followed for data analysis are repeated for each sample. Computers perform such repetitive operations rapidly and virtually without error. This enables the analyst to spend the majority of time interpreting the significance of the results and to spend additional time on the difficult samples, for example, samples containing compounds not identified by library search.

### 4.6.3. Total automated analysis

The ultimate goal of automated GC–MS system development is to limit operator interaction to loading a sample tray and pushing the start key. A combination of advanced software, autosampler operation, and appropriate GC–MS–computer electronic interface devices is required to achieve this objective. The hardware is available now, however, software for complete automated analysis is limited to pre-specified target compounds (qualitative and quantitative) or compounds represented in available mass spectra data compilations (qualitative).

Target compound analysis is ideally suited to automation because the compounds of interest have been previously defined, and optimized GC–MS conditions can be developed. Libraries of mass spectra containing only the desired compounds will enable more rapid data analysis. For most target compound applications pure standards are available, therefore retention times as well as mass spectra can be used for peak indentification. In addition, accurate response factor data can be maintained. Advanced target compound analysis software for completely automated analysis is available now for some commercial GC–MS–computer systems.

Qualitative identification of unknown compounds is much more difficult than target compound analysis. Even if mass spectra give high match factors in library searches, final positive identification should include analysis of a standard solution of the identified substance using the same GC–MS conditions. A useful technique is to co-inject aliquots of the sample and standard solution of the proposed compound. If the identification was correct, then the GC peak shape, retention time, and mass spectra of the unknown will be unaltered, except that the peak will be larger. If poor matches are obtained after using computer search systems, identification must be performed manually using all available data.

Automated identification systems based upon computer interpretation of mass spectra using fragmentation rules have been under development for some time, but with limited results. Interpretation of mass spectra from basic principles of fragmentation requires a much greater understanding of fundamental processes than now exists. It is possible to explain the fragmentation patterns of known structures, but considerably more difficult to attempt the reverse. Good progress has been made in partial identification, such as compound class determination, using mathematical techniques such as pattern recognition. For the forseeable future, totally automated GC–MS–computer systems, from sample injection to final report generation, will be limited to target compound determination or general qualitative scans where most sample components are represented in available reference collections of mass spectra.

### 4.7. GC–MS–computer instrumentation

Application of GC–MS to different fields including environmental, forensic, petrochemical, and the study of disease, has led to the development of many GC–MS techniques and a corresponding increase in instrumental capabilities. Although the array of instrumentation is impressive, most analytical applications are performed using only a few of the available techniques. In order to understand the application of the instrumentation to most problems, it is necessary to have a fundamental knowledge of only the basic components. For many modern instruments, the degree of computerization has become so great that knowledge of software and injection procedures are all that are required for routine analysis.

Fig. 4.32 is a block diagram that shows how the components of a GC–MS–computer system are interrelated. Each system is a complex combination of various physical and electronic elements. The collections of elements that form the gas

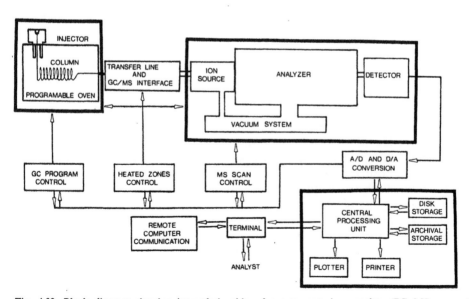

Fig. 4.32. Block diagram showing interrelationship of components in complete GC–MS–computer system.

TABLE 4.10

OPERATIONAL TECHNIQUES IN GC–MS ANALYSIS

| GC–MS analysis component | Available techniques |
|---|---|
| *Gas chromatograph* | |
| Injection technique | Split, splitless, on-column, automatic value |
| Choice of column | WCOT glass, WCOT fused silica, packed |
| Temperature program | Isothermal, linear temperature program, multi-ramp |
| *Mass spectrometer* | |
| Ionization method | Electron impact, chemical ionization, field desorption, field ionization |
| Analyzer | Quadrupole, magnetic, time-of-flight, ion trap, double focussing, MS–MS, FTMS |
| Scan techniques | Mass spectra, selected ion monitor, others related to double focussing and MS–MS |
| *Data system* | |
| Data output | Mass spectra, retention time, peak area, mass chromatogram, reconstructed gas chromatogram, SIM |
| Data analysis | Normalized mass spectra, retention index, library search, quantification, data enhancement (i.e. background subtraction), exact mass determination |
| | *Related techniques* |
| | Autoinjector/automatic analysis Analysis of volatiles by purge/trap Analysis of non-volatiles by pyrolysis GC–MS Cluster analysis Pattern recognition |

chromatograph, mass spectrometer, and computer components are indicated. Other elements of the instrument shown in Fig. 4.32 are related to the interfacing or data transfer between components. The terminal is considered to be a device which interfaces the operator to the system.

Some of the individual elements shown in Fig. 4.32 represent several different techniques that can be employed. For example, various injection techniques and types of analyzers have already been described. Table 4.10 lists many of the techniques available on modern commercially available systems. As hardware capabilities are refined, the difference between various systems with similar features becomes primarily the software. Programs for automatic tuning, instrument operation, data analysis and report generation are already available. The capabilities of a GC–MS–computer system can often be improved more by updating the software than the hardware. In order to see how hardware and software features are brought together in working instruments, it is necessary to examine a few that are commercially available.

### 4.7.1. Finnigan MAT systems

The Finnigan MAT Corporation supplies the greatest diversity of instrumentation of all GC–MS manufacturers. In addition to instruments of general applica-

136

TABLE 4.11

FINNIGAN MAT GC–MS–COMPUTER SYSTEMS

| Product line | Description/application | Special features |
|---|---|---|
| Ion trap detecor (ITD) | Stand-alone GC detector | Unique principle of operation |
| 4500 Series | Quadrupole GC–MS, general applicability | Mass range 4–1800 a.m.u. positive/negative CI on alternate scans upgradable to MS–MS |
| Organics-in-water (OWA) analyzer | Quadrupole MS, designed for analysis of organics in water | Automatic MS tuning, optimized software, head-space sampler and liquid sample concentrator available |
| 8200 Series | Double focussing MS, for high resolution applications and specialized research | Open-split or direct inlet interface, magnet designed for fast scan, special linked-scan techniques available, upgradable to MS–MS (HS Q-30) |
| 8400 Series | Similar to 8200 with extended mass range | Mass range of 8400 a.m.u., resolving power 70,000, improved high mass sensitivity |
| Triple quadrupole (TSQ) | Tandem MS (MS–MS), good for analytical applications where regular GC–MS fails | Special MS–MS scan techniques available |

bility, specialized systems with hardware and software designed for specific applications are produced. Table 4.11 illustrates the range of GC–MS–computer systems by describing a few of them. Several of the systems listed are based upon the quadrupole analyzer which has been discussed in section 4.5.2. The ion trap detector (ITD), however, is interesting because of its method of operation which is unique compared to other types of analyzers described. As an example of high resolution, double-focussing instruments, the Finnigan MAT 8200 will also be described in some detail. Finally, the Finnigan MAT INCOS data system will be used to demonstrate the data acquisition and analysis capabilities of software.

*Ion trap detector.* Fig. 4.33 is a schematic drawing of the Finnigan MAT ITD. It was designed as a stand-alone MS detector, capable of being interfaced to any GC using WCOT columns. The vacuum system is based on a turbomolecular pump, and ions are detected by an electron multiplier. As Fig. 4.33 shows, the ITD is a very simplified system, containing few component parts. Its capabilities are full mass scan for compound identification and operation in SIM mode for quantitation using electron impact (EI) ionization. Other capabilities such as field ionization, field desorption, direct probe operation or other inlet systems such as LC–MS are not available with the ITD. Since the ITD was designed as a universal or selective GC detector for those who are not MS experts, these limitations are appropriate. A

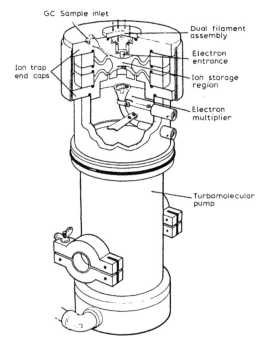

GC Sample inlet

Dual filament
assembly

Electron
entrance

Ion trap
end caps

Ion storage
region

Electron
multiplier

Turbomolecular
pump

Fig. 4.33. Schematic drawing of Finnigan MAT ion trap detector.

further illustration of the simplified design of the ITD is its data system which is based on a personal computer rather than a more powerful minicomputer.

The unique feature of the ITD compared to conventional mass spectrometers is that the ion source and analyzer regions are the same. This is illustrated in Fig. 4.34 which is an expanded, cross sectional drawing of the ITD. Molecules entering the analyzer region are ionized by conventional electron impact. Ions over the entire

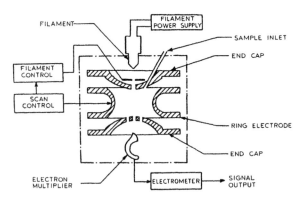

FILAMENT

FILAMENT
POWER SUPPLY

SAMPLE INLET

END CAP

FILAMENT
CONTROL

SCAN
CONTROL

RING ELECTRODE

END CAP

ELECTRON
MULTIPLIER

ELECTROMETER

SIGNAL
OUTPUT

Fig. 4.34. Cross-sectional diagram of Finnigan MAT ion trap detector.

$m/z$ range of interest are not allowed to leave the ionization region, they are trapped by a quadrupole field which is formed by applying a radio-frequency (RF) voltage ($V \cos \Omega t$) between the end cap electrodes and ring electrode. By solving a series of differential equations (Mathieu equation) where $V$ and $\Omega$, and the radius of the ring electrode ($r_0$) are key parameters, it is possible to generate a stability diagram define which $m/z$ values can be trapped using a specified set of conditions.

In the Finnigan MAT ITD design, ($V \cos \Omega t$) is initially set so that all ions of interest above a lower $m/z$ cutoff are trapped during the ionization period of typically 1.0 ms. Several hundred cycles are required after ionization to allow ions of masses lower than the cutoff mass to depart from the trapping field region. By increasing the applied RF voltage ($V$), the lower limit of the range of masses which may be trapped is increased proportionally. Therefore, ramping $V$ will cause stored ions to become unstable in order of increasing $m/z$. Unstable ions will rapidly depart the trapping field region in the direction of the end cap electrodes, and since the lower end cap is perforated, a significant percentage will be transmitted through and are detected by an electron multiplier. It is important to control the sweep rate so ions of consecutive values of $m/z$ are not made unstable at a rate faster than they depart the trapping field region.

Detection limits for the ITD are of the same order as for the flame ionization detector. Although not specifically designed for chemical ionization (CI) analysis, spectra resembling CI may be generated by careful control of operating conditions and addition of appropriate reagent gases. Since negative as well as positive ions may be trapped, negative ion operation is possible.

Mass spectra generated by the ITD are not always identical to those from conventional quadrupole mass spectrometers, although differences are generally not great. For some compounds more abundant molecular ions are formed as in CI. Usually, $(M+1)^+$ ions from addition of $H^+$ are formed. This is not exactly the conventional CI process, but is an ion-molecule reaction which leads to the same product ion. The pressure in the ITD is lower than is used for conventional CI operation. Mass spectral pattern differences from conventional MS could affect the ability to perform library searches for some compounds using external computer matching algorithms such as PBM and STIRS (see section 4.4). A library search algorithm has been developed for the ITD that minimizes differences when searching so that ITD mass spectra can be successfully identified by matching mass spectra from existing data bases.

The real advantage of such simple MS detectors as the Finnigan MAT ITD is that they encourage application of the GC–MS technique to many fields previously reserved for GC using other detectors such as flame ionization and electron capture. Although the full capabilities and detection limits of more complex GC–MS systems are not available, the ITD has more flexibility and greater range of application than any conventional GC detector.

*8200 series GC–MS systems.* The Finnigan MAT 8200 is a high resolution, double focussing instrument. Fig. 4.35 is a photograph of the 8200 showing the gas chromatograph, analyzer and vacuum systems, and electronics panels containing the circuitry for MS tuning and scan control. Although well suited for GC–MS

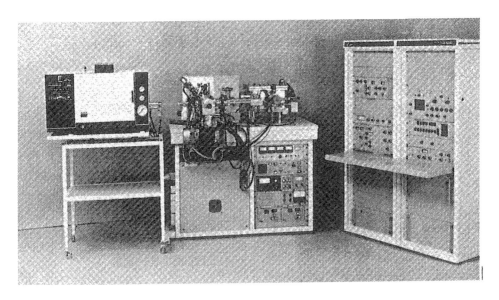

Fig. 4.35. Photograph of Finnigan MAT 8200 double focussing mass spectrometer.

operation, the mass spectrometer can be purchased separately and interfaced to almost any gas chromatograph by an open-split interface or direct connection of a WCOT column to the ion source. It is typical of high resolution systems that they are designed primarily as mass spectrometers, with GC being one of several sample inlet options. This contrasts with most low resolution quadrupole analyzers that are designed as integrated GC–MS systems. In addition to GC, the 8200 can use a direct probe or LC inlet. An automatic direct evaporation system (AUDEVAP) can be used for solid probe, fast atom bombardment (FAB) and direct chemical ionization (DCI) applications. As many as 46 samples can be loaded on a sample magazine and analyzed sequentially by data system control.

A variety of ion sources are available. The standard one is a dual EI–CI source with separate EI and CI chambers for good performance in either mode. An EI-only source can be used that has been optimized for high sensitivity operation. Special techniques such as DCI and FAB are performed using the standard EI–CI source, while another source can be used for field desorption or field ionization analysis.

The 8200 analyzer is a combination of magnetic and electric sectors for double-focussing operation. Fig. 4.36 shows the 8200 ion optical system and reverse Nier-Johnson geometry of the analyzer, in which the magnetic sector preceeds the electric sector. A conversion dynode of 5 kV is used for post-acceleration at the detector to improve detection efficiency at high masses. At full accelerating voltage, masses up to 2100 a.m.u. can be detected. A resolution of 50,000 can be achieved at optimum instrument tuning. Among the most important parameters for high resolution operation are the slit settings. Both entrance and exit slits, shown in Fig. 4.36, are continuously variable and electrically operated. Two resolution settings can

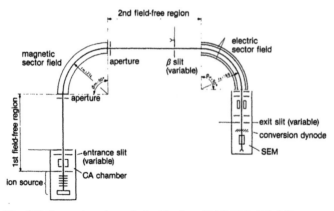

Fig. 4.36. Ion optics schematic for Finnigan MAT 8200 GC–MS system.

be pre-set and selected by push-button. The slits can also be rotated to ensure that they are parallel to the ion beam.

Fig. 4.37 illustrates the operation of the 8200 at a resolution of 50,000. Two ions differing in mass by only 0.00447 a.m.u. are distinguished at this resolving power. Elemental compositions of ions can be determined at this resolution by exact mass measurement techniques. These generally involve comparing unknown peaks with known masses of peaks from a calibration compound that is continuously bled into

$m/z = 224$, R = 50,000 resolving power

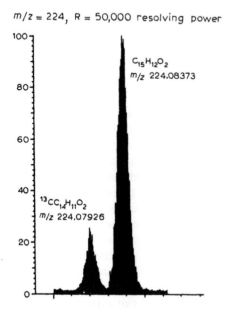

Fig. 4.37. Operation of Finigan MAT 8200 at 50,000 resolution.

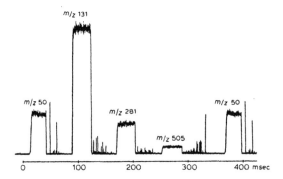

Switching between 4 ions in one mass decade

Fig. 4.38. Operation of Finnigan MAT 8200 in SIM mode.

the ion source during the analysis. For quantitative determinations by SIM, high resolving power permits the analyzer to eliminate peaks having $m/z$ close to those of analyte peaks that otherwise would interfere with low resolution MS analysis. When performing analytical work the resolution should be set at the minimum required to eliminate interferences. This provides maximum sensitivity, since higher resolution settings require narrow slits and reduce the number of ions reaching the detector.

Recent designs of double focussing instruments give greatly improved scan speeds. This has been made possible primarily by the use of laminated magnets that have very fast recovery times. Previously, magnets could not be scanned rapidly and then reset for the next scan without experiencing hysteresis effects. Times between scans to stabilize the magnet were great enough to preclude high resolution operation on fast eluting GC peaks from WCOT columns. Such problems affected multiple-ion SIM operation as well, since a lengthy setting time was required after the magnet was "jumped" from one mass value to the next, unless the two masses were close together. Both improved electronics and magnet design have greatly reduced this problem.

The performance is illustrated in Fig. 4.38. Switching between four masses in one decade (i.e. $m/z$ 50–500) is shown, requiring a total jumping time of about 200 ms. The same switching speed can be maintained over a large mass range up to $m/z$ 1000. This is adequate for multiple ion SIM analysis at low to medium resolution (up to ca. 10000) for the narrow GC peaks of WCOT columns. For applications of SIM where the elimination of interferences is a primary consideration, resolutions of 7000–10,000 are generally sufficient.

Fast scan rates can also be used to obtain full mass spectra at low to medium resolution. The Finnigan MAT 8200 can operate routinely at scan rates of 0.3 s/decade, which includes a 0.2-s scan and 0.1 s recovery time. A trade-off has to be made between scan speed and resolution since high resolution operation requires slower scan speeds. For WCOT column GC analysis, typical scan rates up to a resolution of ca. 10,000 are 1–2 s/decade. Mass measurements in single scans at a

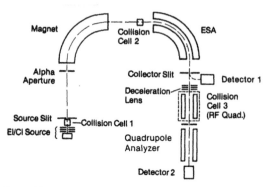

Fig. 4.39. Ion optics schematic for 8200 series MS–MS.

resolution of 5000 and scan speed of 1 s/decade for the 8200 are accurate to ca. 0.0013 a.m.u.. Accuracy of the measurement can be improved by taking the average of several scans. Instrument operation at resolution values greater than 10,000 require scan speeds too slow for use with WCOT columns.

The full capabilities of the Finnigan MAT 8200 as an organic reaction and biochemical research tool extend past the scope of basic analytical GC–MS. Because of its reverse Nier-Johnson geometry, metastable ion transitions can be analyzed in both field free regions. A linked-scan microprocessor unit allows parent ion, daughter ion, and constant neutral loss scans to be obtained for these transitions. Techniques such as mass analyzed ion kinetic energy spectroscopy (MIKES) can be employed in these studies. To extend the capabilities of the 8200 further, a quadrupole analyzer can be added after the electrostatic sector to form a triple analyzer MS–MS system. The ion optics of this system is shown in Fig. 4.39. Such a system, in addition to the abilities already described, has the features of MS–MS analysis described in Chapter 5.

*INCOS data system.* The computer programs that comprise the INCOS data system are one of the most successful sets of software now in use. INCOS was designed to control instrument operation, acquire and store GC–MS data, perform data analysis, and report results. Versions of INCOS are available for both magnetic and quadrupole instruments, and the data system can be interfaced to most GC–MS systems. It is beyond the scope of this book to completely describe the capabilities of any advanced GC-MS software, so only a general summary of INCOS is presented. The purpose of this description is to provide an understanding of the many considerations of an advanced GC-MS data system, not only for data acquisition and analysis, but also the many support programs that are necessary.

INCOS exists in two principal programming levels. The most basic of these is the IDOS command processor (INCOS Disk Operating System), which is necessary for any other programs to run on the Data General Nova series minicomputers upon which INCOS is based. Programs and commands to operate the GC–MS and to acquire and analyze data run under the Mass Spectrometer Data System (MSDS) command processor. INCOS consists of both IDOS and MSDS. Only MSDS will be

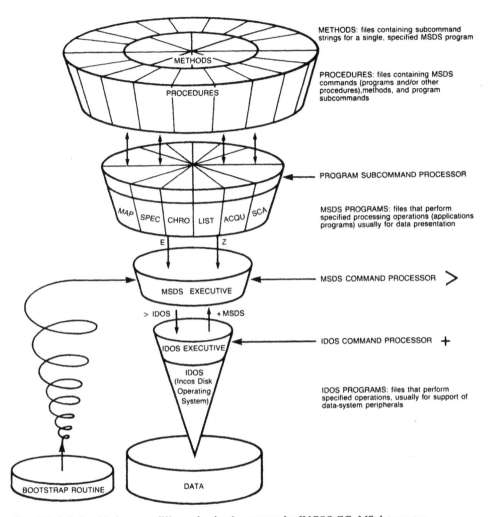

Fig. 4.40. Relationship between different levels of programs for INCOS GC–MS data system.

discussed here. The relationship between different programming levels is illustrated in Fig. 4.40. When the computer is initially turned on, a small program is used to initialize the system (bootstrap routine) and to allow access to the GC–MS operating programs through the MSDS Executive. All MSDS programs can be called from the Executive by typing their name on a terminal and pressing the carriage return key (CR). When running various programs, important GC–MS variables are saved in a status table, that can be displayed by entering STAT(CR).

Several MSDS programs are available, some of which appear in Fig. 4.40 (MAP, SPEC, CHRO). Each of these programs has a series of subcommands that are executed through the program subcommand processor. In addition to the levels of programs already described, users can write their own special routines by connect-

TABLE 4.12
DESCRIPTION OF PRINCIPAL MSDS PROGRAMS

*A. Instrument control and data acquisition programs*

| | |
|---|---|
| SYST | Creates instrument descriptor file that stores key parameters used by the mass spectrometer to acquire data. |
| SCAN | Defines mass range and time spent on each scan. |
| MID | Creates SIM descriptors containing a list of ions monitored, dwell times and other relevant information. |
| ACQU | Used to acquire data in either full scan or SIM mode. ACQU specifies how much data is acquired, when it is started, and can start the GC run and turn on/off the ion source filament and detector. |
| DESC | Creates a GC descriptor containing parameters for temperature programming operation. |
| SAMP | used for control of autosampler. |

*B. Data display programs*

| | |
|---|---|
| MAP | Used for real-time or post-run display of acquired full scan or SIM data. |
| CHRO | Similar to MAP but only data already acquired can be displayed. CHRO is also used for quantitative analysis. |
| SPEC | Displays single scan data. HSPE used for high resolution data. |
| LIST | Lists masses and intensities for a single scan. HLIS used for high resolution data. |
| COMP | Assigns possible elemental compositions to ions based upon accurate mass data. |

*C. Data processing programs*

| | |
|---|---|
| CALI | Calibrates mass analyzer using data obtained by scanning a reference compound (usually perfluorotributylamine). |
| QUAN | Used for quantitative data analysis. |
| LIBR | Forward library search. |
| SEAR | Reverse library search. |

*D. Utility programs*

| | |
|---|---|
| STAT | Displays current MSDS and GC-MS status. |
| FILE | Used to list, transfer, or delete files. |
| EDNL | Edit name list - used for automatic data processing, creates list of data files specified by user. |
| EDQL | Similar to EDNL, for automated quantitation. |
| EDLL | Similar to EDNL, for automated library search. |

ing several MSDS commands using a Procedure language. Such procedures may also contain files that have subcommand strings for specific MSDS programs, called Methods. A large amount of routine data analysis may be automatically performed by Procedures and Methods. Very little programming knowledge is required, one needs only to be familiar with the available MSDS commands and subcommands. For data analysis requiring more complex programming, the Fortran language can be used. All of the various MSDS programs are written in Fortran, and it is possible for them to be tailored for specific applications.

The principal MSDS programs are described in Table 4.12. All of these programs have many subcommands associated with them, which can be displayed by typing the program name followed by a question mark. For example, Fig. 4.41 was generated by the command ACQU?/H (CR). The (/H) option instructed the

```
ACQU accepts the following subcommands:                        03/23/83
D  (H)         Display (hardcopy) status
G  (*,<,>) J       Start a J scan aquisition.  Write a log file unless J=1
G  (*,<,>) J:k:l Start a J hour,  k minute, l second acquisition.
               (*=delete any log file and write none)
               (<=always make a new log file) (>=append to the log file)
G' J           Start a J cycle acquisition of current experiment
A  (aa) (ac)      Ask accounting questions (update header) (ask conditions)
X              Make an entry in the log file for the current acquisition
L              Toggle state of sensor device acquisition
C              Toggle state of centroid/profile flag
N              Toggle state of positive/negative ion flag
R  (R1)        Toggle state of RIC-only flag, R1 toggles temperature flag
S              Toggle state of suspend flag
   J           Change number of scans to J
   J:k:l       Change total acquisition time to J:k:l
   O           Stop the acquisition at end of current scan
W  J           Set minimum peak width to J centroider samples
F  J           Set minimum fragment width to J % of peak width
T  J           Set ADC threshold to J
M  J           Set minimum area to J
B  J           Set baseline to J
I  (K)         Initialize (keep) acquisition parameters
Q              Hardcopy this page
E  (Z)(ID)     Exit to MSDS ( to MSDS ) ( to IDOS )
J  J(:k:l),n
               Wait maximum n seconds for sample injection,
               then execute "O" subcommand for J scans (h:m:s)
O  m           Start GC.  If m>0, go execute "G" command, m=#scans
#XY (*XY)      Load GC-descriptor XY into the GC ( into display )
&XY (%XY)      Start MID-scan ( experiment ) from descriptor XY
@XY            Execute commands from file ACQUXY.ME
!x             Replace x with the value of the  MSDS variable (x=10-19)
ACQU accepts the modifier /O to acquire even if the run name exists
```

Fig. 4.41. Description of subcommands available with the MSDS program ACQU. Knowledge of these commands is needed to acquire data when using the INCOS GC–MS Data System.

computer to print the list of subcommands for the program ACQU. Fig. 4.41 shows that there are many options available. Some experience is required to set these parameters so that data acquisition is optimized, although most parameters have default values set to enable efficient GC–MS operation even if all of the subcommands are not understood by the user.

The use of MSDS programs and subcommands to develop special defined Methods and Procedures is illustrated in Fig. 4.42, which is the listing of a Procedure called PCDD45. This procedure controls the SIM analysis of samples for chlorinated dibenzo-*p*-dioxins and dibenzofurans using five different groups of ions containing six ions each. A total of thirty *m/z* values were needed to monitor all the compounds of interest, but putting all thirty together in a single group would have resulted in poor detection limits. The thirty ions could be split into five groups because the compounds monitored by each group are completely separated from each other by their GC retention times. An alternative would be to use only one group of six ions at a time, and inject the sample five times. An explanation of how PCDD45 works is shown in Table 4.13.

Similar procedures to that described above can be developed for automated data

146

```
TRACE OF PROCEDURE PCDD45
    * ACQU/O  (&L4; G450; E)
    * MAP/C(@M4; E)
    * ACQU  (&B5; E)
    * MAP(@M5; E)
    * ACQU(&B6; E)
    * MAP(@M6; E)
    * ACQU(&B7; E)
    * MAP(@M7; E)
    * ACQU(&B8; E)
    * MAP(@M8; E)
    * STOP
    *
```

Fig. 4.42. Procedure for acquired data in SIM mode for the analysis of samples for chlorinated dibenzo-*p*-dioxins and dibenzofurans.

output, quantitation, or library search. To perform such operations on a series of data files, a namelist is first created with EDNL, EDQL or EDLL (see Table 4.12), which contains all the names of data files to be analyzed. Files can be sequentially called from the list by simple commands in a manner resembling the use of do-loops in Fortran. Individual files are then processed by a series of MSDS commands and subcommands as shown for the above Procedure.

TABLE 4.13
EXAMPLE OF STEPS USED TO ACQUIRE DATA FOR INCOS SYSTEM

---

*ACQU/O (&L4; G450; E)*

| | |
|---|---|
| ACQU | Calls the program ACQU (see Table 4.12). Terms in brackets are subcommands, and must be separated by semicolons. |
| /O | Tells the program to store data in the (previously defined) file name, even if the file already contains data. |
| &L4 | Loads the SIM descriptor L4 for data acquisition. L4 is a Method that contains SIM parameters such as $m/z$ values, dwell times, and other related information. |
| G450 | Get 450 scans. For SIM operation, a scan is one cycle of consecutively monitoring each of the chosen $m/z$ values. Total acquisition time can be specified instead of number of scans. |
| E | Exit the program after acquiring 450 scans. |

*MAP/C (@M4;E)*

| | |
|---|---|
| MAP/C | Used for real-time display of ion abundances on a CRT. Data are not displayed until acquisition has begun (/C). Terms in brackets are MAP subcommands. |
| @M4 | The method M4 determines which ions are displayed. M4 contains the following subcommands: |

*A∅; 322; 334; 306; D12:40; 14:25*

| | |
|---|---|
| A∅ | Controls the display format |
| 322 | One of three $m/z$ values to be displayed. These masses must be contained in SIM descriptor L4. |
| D | Display subcommand. Abundances of the three ions (322, 334, 306) will be plotted on a CRT between retention times 12:40 and 14:25 min:s. Scan numbers can be specified instead of times. |
| E | After retention time 14:25, exit MAP and load the next group of ions to be monitored [ACQU (&B5;E)]. |
| STOP | Stop acquisition. |

---

An important feature of INCOS is that data analysis commands and Procedures can be executed at the same time as the GC–MS system is analyzing a sample. This is possible because INCOS is a foreground/ background operating system. Data acquisition is always done in foreground mode, and has top priority. Since this only requires 10–20% of the total computer time, the remaining time can be spent in displaying or analyzing data that have already been acquired. For example, as a GC peak elutes and is detected it can be made to appear on a CRT display by using the program MAP. Using the same terminal, the operator can exit MAP, obtain the mass spectrum of the peak, perform a library search, print the best matches, and return to MAP to observe how the acquisition is progressing. Foreground/background operation ensures that maximum use of the GC–MS data system is obtained.

*4.7.2. Hewlett-Packard systems*

The Hewlett-Packard GC–MS instruments are all based on use of the quadrupole mass analyzer. In the early 1970s these mass analyzers were called dodecapole because the circular mass analyzer rods used had trimmer electrodes to simulate the hyperbolic fields required by theory. Today, Hewlett-Packard quadrupole mass analyzers are constructed with highly precise hyperbolic shaped rods to provide the proper fields for optimum separation and transmission of ions of high as well as low masses. The performance of the components of quadrupole GC–MS systems of different manufacturers and designs are fundamentally comparable. The greatest differences are in the flexibility and power of the associated computer and its software and in the way different components of the basic instrumentation and computer elements are assembled. Different systems range from simple with limited capabilities to complex with maximum capabilities for both data acquisition and analysis.

There are three different Hewlett-Packard GC–MS systems each designed to provide capabilities for a specific level of analytical scope. The largest and most versatile is the HP5988. It can perform all the necessary GC–MS functions and has a powerful computer and software which allows simultaneous multitasking and multiuser activities. The HP5995 is a benchtop GC–MS system that can perform many of the GC–MS functions, but on a more limited scale. The HP5970 is a GC–MS system, named Mass Selective Detector, that is essentially dedicated to capillary column, electron ionization use. Although it can be obtained with an integrated HP gas chromatograph it can also be used as a highly selective detector with any gas chromatograph. Each of the three major systems described above all use the same quadrupole mass analyzer and are modular in design so that different components of hardware and software can be assembled to produce a GC–MS system with specific capabilities.

*HP5988 system.* This is a powerful and versatile system. The instrumental components are those that have been in HP systems for over ten years, with modifications that experience has shown to be necessary for optimum performance and reliability. The block diagram outlined in Fig. 4.43 shows all the provisions needed to produce both EI and CI spectra using separate and high capacity vacuum

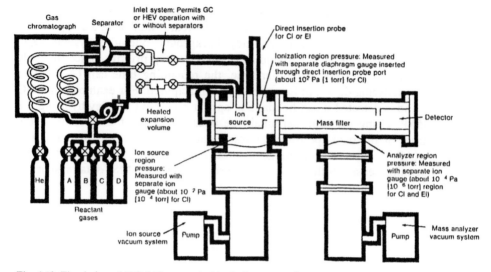

Fig. 4.43. The design of GC–MS system in block diagram outline.

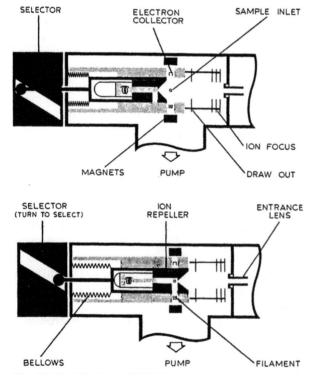

Fig. 4.44. Details of the HP5988 ion source to indicate method to actuate EI and CI modes externally.

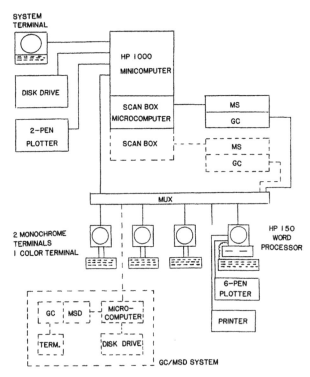

SYSTEM
TERMINAL

HP 1000
MINICOMPUTER

DISK DRIVE

SCAN BOX
MICROCOMPUTER

MS

GC

2-PEN
PLOTTER

SCAN BOX

MS

GC

MUX

2 MONOCHROME
TERMINALS
1 COLOR TERMINAL

HP 150
WORD
PROCESSOR

6-PEN
PLOTTER

GC | MSD

MICRO-
COMPUTER

PRINTER

TERM.

DISK DRIVE

GC/MSD SYSTEM

Fig. 4.45. Functional diagram of an HP5988 computer system and related GC–MS instrumentation.

pumps for both the ion source and the mass analyzer. The ion source design that permits rapid change-over from EI to CI operation by actuation of a single external lever is shown in Fig. 4.44. Although packed GC columns can be used, the system is optimized for capillary GC columns for which the most feasible GC–MS interfaces are the direct inlet and the open-split.

The major advance in the HP5988 system lies in the powerful computer and software it uses. The functional diagram shown in Fig. 4.45 illustrates the interflow of control and information between the major components of the computer system and the GC–MS instrumentation of a system shown in Fig. 4.46. The heart of this system is the HP1000 computer with a disk drive memory of 132 Mbytes that can also utilize a tape drive of 67 Mbytes for long term data storage. The scanning interface is really a separate, "outboard" microcomputer that controls tuning parameters of the quadrupole mass analyzer, the mass scanning, ion peak detection, initial data acquisition, temporary storage and then transfer to the large disk memory. Use of the scanning interface computer in this way frees the large central computer to do other tasks that may be required by any one of the six graphics terminals that can simultaneously access the system or the printers and plotters interfaced via the multiplexer.

All the operating and analysis parameters of the gas chromatograph can be initialized and controlled by the central computer. The gas chromatograph can also

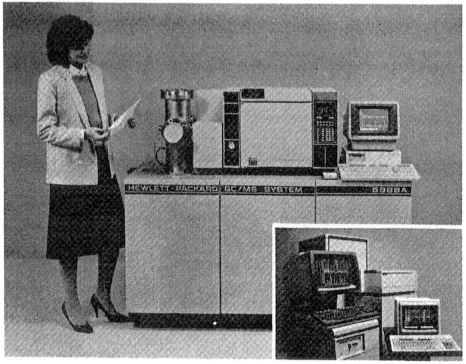

Fig. 4.46. A typical HP5988 system showing the GC–MS unit with the system terminal. The insert shows the computer and operator terminals.

be operated as a stand alone unit and the status of its parameters can be examined and gas flows adjusted independent of the data system.

*Tuning for mass calibration.* Adjusting a quadrupole mass analyzer to obtain the proper sensitivity, resolution and mass scale calibration throughout its mass scanning range can be a very difficult task to do manually. At least seven electrical parameters must be adjusted iteratively until the desired performance is obtained. The hyperbolic rod design makes it possible to allow the computer to do this automatically using a software algorithm called AUTOTUNE. After introduction of the standard perfluorotributylamine (PFTBA), the software automatically and iteratively adjusts the quadrupole parameters to converge on a target set of mass spectral conditions for the PFTBA standard. Once the target parameters are reached a print-out report is given and these conditions are then used for subsequent mass spectra produced. Typical results are shown in Fig. 4.47.

Unlike EI, positive CI cannot be automatically tuned and therefore the mass spectrometer must be manually tuned. It is best to start with the parameters as set by the AUTOTUNE program used for EI tuning and make the necessary changes in the various parameters during the tuning procedure. The electron energy is initially set at 230 eV and the focussing lens assembly is given a potential of 130 V. These

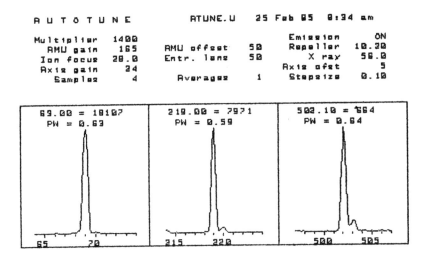

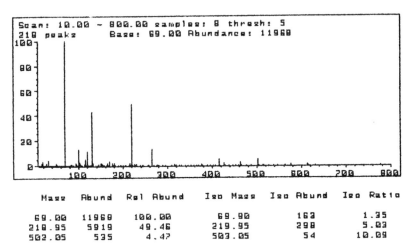

Fig. 4.47. Printout of results for an AUTOTUNE procedure used to adjust parameters for the Hewlett-Packard hyperbolic rod quadrupole mass analyzer.

parameters may require adjustment later in the manual tuning of the spectrometer, however they are set at these values to provide a reasonable starting point.

*Positive chemical ionization.* For positive chemical ionization (PCI) using methane as the reagent gas, the source temperature is typically set to 200°C. Ultrahigh purity methane is introduced into the ion source and the source pressure is adjusted to give the maximum abundance of the $CH_5^+$, $C_2H_5^+$, and $C_3H_5^+$ ions appearing in the methane mass spectrum. Benzophenone (mol. wt. 182) is then introduced into the mass spectrometer via a calibration probe. The $[M + H]^+$ ion in the methane PCI mass spectrum of benzophenone is scanned while the various potentials are manually adjusted to give the maximum intensity of the $m/z$ 183 ion. PFTBA (671) from another sidearm of the probe is then allowed to enter the ion source and the mass axis is calibrated using the $[M + H]^+$ ion of benzophenone and two ions from the PCI mass spectrum of PFTBA, $[M - C_4F_{11}]^+$ and $[M - F]^+$ ($m/z$ 414 and 652, respectively). The manually created tuning file is stored in the data system and can be used for setting the mass spectrometric parameters for methane PCI–MS analyses.

*Negative chemical ionization.* In tuning the HP5988 GC–MS system for negative chemical ionization operation using methane as the moderating gas, the system is first tuned for methane PCI. The source temperature is reduced to 100°C or 150°C (the higher temperature gives lower sensitivity but reduces source contamination). To observe negative ions, the X-ray voltage is reduced to 5 V and the polarity of the mass spectrometer optics are reversed with respect to PCI operation. The mass spectrometer is manually tuned while monitoring the $M^-$ ion of benzophenone ($m/z$ 182). PFTBA is then reintroduced into the ion source and the mass axis is calibrated using the ions $[M - C_4F_{11}]^-$ and $[M - F_2]^-$ ($m/z$ 414 and 633) from PFTBA and $m/z$ 182 from benzophenone. The manually created tuning file can then be stored in the data system as was done for the PCI manual tune.

*Other Hewlett-Packard Systems.* The simpler GC–MS systems of Hewlett-Packard are designed for those whose applications do not require the more versatile, multi-tasking capabilities of the large system. The HP5995 bench top system shown in Fig. 4.48 can be used for either packed or capillary column, or a direct insertion probe, operation. A membrane interface for packed columns, and either an open-split or direct inlet interface for capillary columns is used. The major differences between this system and the larger HP5988 is that less versatile software and a smaller vacuum system is used. This limits the mass spectral search functions and permits only EI operation. The HP5995 GC–MS system can be interfaced to the computer system of the HP5988 and all the expanded capabilities of it are then useable.

The simplest GC–MS system is formed by adapting the HP5970 Mass Selective Detector (MSD) to a gas chromatograph as seen in Fig. 4.49. The MSD utilizes a turbomolecular pump for a vacuum system and can be interfaced to a gas chromatograph for dedicated capillary column, EI use. The software for this system can be provided either from the large HP1000 RTE system used for the HP5988 or from the smaller computer systems used for the bench-top GC–MS.

*The RPN data processing system.* The mass spectral data processing system supplied with Hewlett-Packard GC–MS systems is called RPN, which is an acronym for "reverse polish notation". The system is structured to manipulate and display GC–MS data in a manner similar to an HP hand-held calculator.

RPN consists of a stack of data registers and commands which will manipulate the data and perform utility functions. The stack contains four registers, labelled X, Y, Z and T, each of which can contain an entire chromatogram, mass spectrum or alphanumeric information. The most recently entered data reside in the X-register while the old data are pushed up into the next higher register. This structure allows for very complex manipulations to be performed with a small number of simple

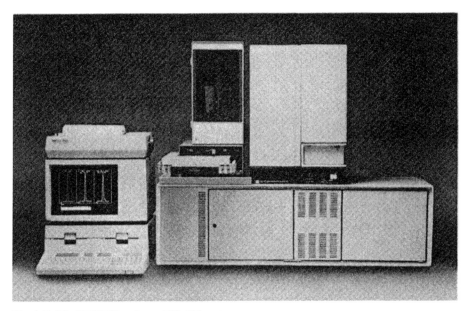

Fig. 4.48. The HP5995 bench-top GC–MS system.

commands. The structure of the stack and the results of the sequence of RPN commands which produce a background-subtracted mass spectrum are presented in Fig. 4.50.

Each of the over one hundred RPN commands is a separate program that can be executed by typing its name into the data system. The commands can be grouped by

Fig. 4.49. The HP5970 mass selective detector interfaced with an HP5890 gas chromatograph.

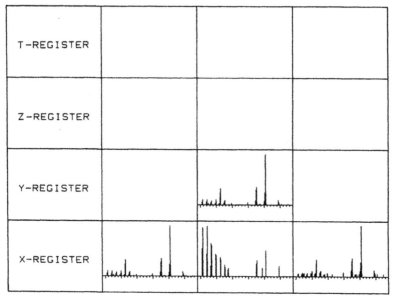

Fig. 4.50. The Hewlett-Packard data system uses a reverse polish notation (RPN). It includes a stack of 4 registers in which to manipulate data. The use of these registers to produce a background-subtracted spectrum is shown.

function into four types: utility, register manipulation, data manipulation and data display. Register manipulation commands are used to move the contents of the stack around prior to data analysis. Data manipulation includes extracting chro-

| Commands | Explanation |
|---|---|
| `:RU,DR,4,,ALL,,,23B,31463B` | ; Draw the RGC on the<br>; terminal screen. |
| `:DP,CHOOSE THE WINDOW OF INTEREST,`<br>`:DP,MOVE THE CURSOR TO THE LOWER LEFT CORNER,`<br>`:DP,STRIKE ANY KEY,`<br>`:DP,MOVE THE CURSOR TO THE UPPER RIGHT CORNER,`<br>`:DP,STRIKE ANY KEY.` | ; Instruct the user to<br>; select a portion of<br>; the RGC using the<br>; cursor. |
| `:RU,LP,2` | ; Turn the cursor on and<br>; capture the (x,y)<br>; values of the corners. |
| `:RU,EC,,,OR:2R` | ; Generate an RGC within<br>; the selected retention<br>; time window. |
| `:RU,NEW` | ; Clear the graphics from<br>; the terminal screen. |
| `:RU,DR,4,,OR:2R,,,23B,31463B,1R:3R` | ; Draw the selected<br>; portion of the RGC on<br>; the terminal screen. |

Fig. 4.51. This sequence of commands creates a procedure file in the RPN system to custom design a data analysis procedure that will automatically execute.

```
:,MANUAL
:**-------------------------------------------------------------
:**
:** MANUAL:PROGRAM FOR QUALITATIVE ANALYSIS OF RUNS
:** LCD 850506
:**
:**-------------------------------------------------------------
:**

   ENTER THE FILE NAME :>Y1458

  reg type  # pts  scan#    range: amu\r.t.      base   file      ion range
  --- ----  -----  -----   ----------------    -------  -----   ---------------
   X   GC   3144     1       5.18-    55.03    92928.0 >Y1458   50.00- 500.00
  reg type  # pts  scan#    range: amu\r.t.      base   file      ion range
  --- ----  -----  -----   ----------------    -------  -----   ---------------
   X   GC   3144     1       5.18-    55.03    92928.0 >Y1458   50.00- 500.00


MANUAL; Select a softkey....



: TR,PKSRCH::P1
  reg type  # pts  scan#    range: amu\r.t.      base   file      ion range
  --- ----  -----  -----   ----------------    -------  -----   ---------------
   X   GC   3144     1       5.18-    55.03    92928.0 >Y1458   50.00- 500.00
  working...
  draw complete....

CHOOSE THE WINDOW OF INTEREST BY
MOVING THE CURSOR TO THE LOWER LEFT CORNER AND HITTING ANY KEY THEN
MOVING THE CURSOR TO THE UPPER RIGHT CORNER AND HITTING ANY KEY
  reg type  # pts  scan#    range: amu\r.t.      base   file      ion range
  --- ----  -----  -----   ----------------    -------  -----   ---------------
   X   GC   1142    889     19.28-    37.38    31120.0 >Y1458   50.00- 500.00
  working...
  draw complete...

  WINDOW OK? (Y/N) : Y


  ENTER DETECTION THRESHOLD :.22


  ENTER PERCENT AREA THRESHOLD : 8


>Y1458        LF-7-11-4          LF-7-11-4(20uL)EILIN
  50.0| 500.0 TIC

  Peak   R.T.  first  max   last   peak    raw     corr.    corr.    % of
   #     min.  scan   scan  scan   height  area    area     % max.   total
  ---   ------ -----  ----- -----  ------  -------  -------  -------  -------
   1    22.249  1065  1076  1082   12999   164362   118762   45.44   15.163
   2    24.233  1189  1201  1216   28483   329169   261345  100.00   33.367
   3    26.406  1328  1338  1349   16156   199340   133680   51.15   17.067
   4    30.196  1571  1577  1585    3640    44326    22999    8.80    2.936
   5    33.683  1789  1797  1808    9114    82393    61385   23.49    7.837

   6    34.380  1833  1841  1848    3455    43199    27289   10.44    3.484
   7    36.868  1992  1998  2009   28828   178557   157788   60.38   20.145

                       Sum of corrected areas:      783248.
  working...
  draw complete...

  INTEGRATION OK? (Y/N)  : Y


  SUBTRACT BACKGROUNDS? (Y/N)  : Y
```

Fig. 4.52. An automated quantitative analysis procedure illustrated for the RPN system.

matograms or spectra from GC–MS data and placing them on the stack, adding or subtracting two spectra and integrating a chromatogram. The data display commands provide flexibility in presenting analytical results in a readily understandable form. The analyst can select the format of the graphics, label plots and route the output to terminal, plotter or printer. Various utility commands give information on specific data files and RPN system status, perform statistical analyses and manipulate system resources.

```
1. Phenol, pentachloro-                          264  C6HC15O
2. Phenol, pentachloro-                          264  C6HC15O
3. Phenol, pentachloro-                          264  C6HC15O
```

```
Sample file: >Y1458    Spectrum #:        1076
Search speed: 1        Tilting option: S      No. of ion ranges searched:    60
```

|      | Prob. | CAS # | CON # | ROOT | K | DK | #FLG | TILT | % | CON | C_I | R_IV |
|------|-------|-------|-------|------|---|----|------|------|---|-----|-----|------|
| 1.   | 97*   | 87865 | 59841 | "BIGDB | 152 | 25 | 1 | 0 | 84 | 21 | 53 | 99 |
| 2.   | 95*   | 87865 | 59842 | "BIGDB | 100 | 78 | 0 | 0 | 76 | 21 | 53 | 97 |
| 3.   | 86*   | 87865 | 59845 | "BIGDB | 115 | 38 | 2 | 2 | 82 | 7 | 59 | 76 |

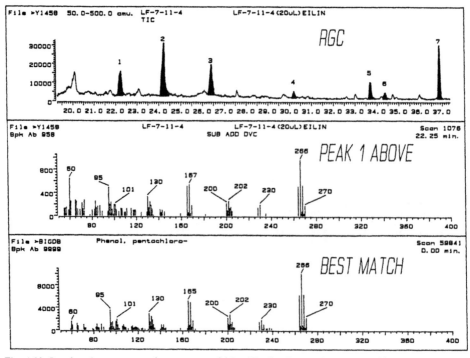

Fig. 4.53. Results of an automated mass spectral identification by a reference matching procedure. The value of 97 for probability of compound 1 gives high assurance to identification.

An important feature of RPN is the ability to write "procedure files", which contain RPN and FMGR commands that execute automatically when control is transferred to that file. To illustrate, Fig. 4.51 presents an extracted portion of a procedure file. The RPN commands are preceeded by RU (Run) since they are programs to be executed by the computer. The DP commands are FMGR commands so they need not be preceeded by RU. The function of this set of lines is to display a reconstructed gas chromatogram (RGC), ask the operator to select a portion of the RGC by defining diagonally-opposite corners of the "window of

```
1. 7H-Benz[de]anthracen-7-one                          230  C17H10O


Sample file: >J1810      Spectrum #:       2385
Search speed: 1          Tilting option: S    No. of ion ranges searched:   60


        Prob.    CAS #   CON #   ROOT     K    DK  #FLG TILT  %   CON  C_I R_IV

  1.     30*     82053   54197   "BIGDB   28   104   3   0   80    32   12   13
```

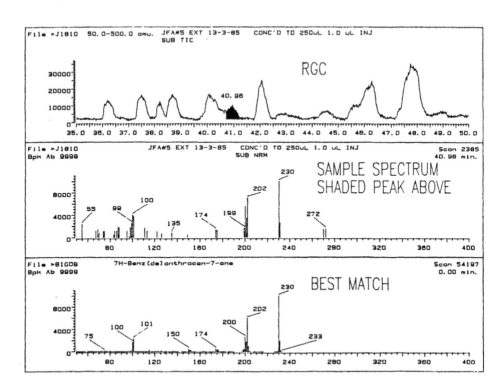

Fig. 4.54. When a poor match, as seen by a probability of 30, occurs using the automated procedure, then a manual search procedure must be used.

interest" and drawing that portion on the terminal screen. Using procedure files, "custom-designed" data analysis procedures can be written without a detailed knowledge of programming.

*Automated search routines with Hewlett-Packard GC–MS system.* The identification of unknown mass spectra using computer searches of a spectrum library can be a lengthy operation. Fortunately procedure files can be written to automate this process. One example of this is the operator-written "Auto Search" routine in the procedure file "MANUAL". In response to a prompt from the procedure, the operator enters the sample file name. Once the data file is verified the operator can access the "Auto Search" routine. The RGC is drawn on the terminal screen and the operator is asked to select a portion of the RGC by moving a cross-hair cursor to define the lower-left and upper-right corner of the area of interest. This area is drawn on the screen, replacing the other RGC plot. The procedure asks if the drawn portion is the correct one, and if not, gives the operator a chance to try again. The portion of the RGC is integrated to find the peaks which are then indicated on the RGC trace on the screen. The operator can change integration parameters and view the results until the peaks of interest have been selected. From this point, the procedure can run unattended.

Each peak is background-subtracted before the PBM search, the average of the first and last spectra of the integrated peak being taken as the background. The PBM search results for each peak are printed along with the RGC, the sample spectrum and the best matching spectrum from the spectra library. Portions of the dialogue with the procedure are shown in Fig. 4.52 and 4.53.

*Other search facilities.* Often a PBM search of a peak spectrum will not yield a clear-cut identification, as illustrated in Fig. 4.54. A poor match could be caused by a co-eluting compound or a high noise level. Generally, when computer searching fails to yield an acceptable match, the analyst is forced to resort to manual interpretation. Procedure files can be written to provide additional tools to aid in identifying unknown components in a mixture.

The "Manual Search" routine of the "MANUAL" procedure file provides this facility. Once an area of interest in the RGC has been chosen and displayed, the routine presents several options. A mass spectrum can be chosen by using the cursor. Prior to a PBM search, the cursor can be used to choose a background spectrum which may be different than that chosen by the automatic search. Upon examination the operator may suspect a peak to contain co-eluting compounds. The routine will plot a set of chosen mass chromatograms. If there are MC peaks appearing that do not have the same retention time, the original peak does contain co-eluting compounds. By the judicious choice of spectra across the peak the different compounds can be identified.

## 4.8. Suggested reading

1 W. McFadden, *Techniques of Combined Gas Chromatography/Mass Spectrometry: Applications in Organic Analysis*, Wiley, New York, 1973.
2 B.S. Middleditch (Editor), *Practical Mass Spectrometry, A Contemporary Introduction*, Plenum Press, New York, 1979.

3 A.M. Greenway and C.F. Simpson, *J. Phys.*, 13 (1980) 1131–1147.

4 F.W. Karasek and A.C. Viau, *J. Chem. Educ.*, 61 (1984) A233–236.

5 B.J. Guizinowicz, M.J. Guzinowicz and H.F. Martin, *Fundamentals of Integrated GC – MS, Part III: The Integrated GC – MS Analytical System (Chromatographic Series Vol. 7)*, Marcel Dekker, New York, 1977.

6 J.E. Biller and K. Biemann, *Anal. Lett.*, 7 (1974) 515–528.

7 L. Van Vaeck and K. Van Cauwenberghe, *Anal. Lett.*, 10 (1977) 467–482.

8 F.W. Karasek, *Res./Develop.*, 27 (November) (1976) 42–46.

9 G.C. Stafford, Jr., P.E. Kelley and D.C. Bradford, *Amer. Lab.*, June (1983).

10 G.C. Stafford, Jr., P.E. Kelley, J.E.P. Syka, W.E. Reynolds and J.F.J. Todd, *Int. J. Mass Spectrom. Ion Proc.*, 60 (1984) 85–98.

11 W.T. Wipke, S.R. Heller, R.J. Feldman and E. Hyde, *Computer Representation and Manipulation of Chemical Ionization*, Wiley, New York, 1974.

12 G.M. Pesyna and F.W. McLafferty, in F.C. Nachod, J.J. Zucherman and E.W. Randall (Editors) *Determination of Organic Structures by Physical Methods*, Vol. 6, Academic Press, New York, 1976, Ch. 2.

CHAPTER 5

# GC–MS–COMPUTER APPLICATIONS

The use of GC–MS–computer techniques in a wide range of applications has advanced our understanding of chemical systems at an incredible rate. Improved techniques which are pushing detection limits below the ppt level and improving the ability to identify unknown substances insure that the applicability of GC–MS–computer to solving complex problems will continue to increase in the forseeable future. Our recent understanding of the existence of trace levels of toxic organic compounds in the environment would not have been possible without GC–MS–computer methods, and some truly elegant applications of GC–MS have been reported. Equally impressive advances in many other fields including forensic and the study of disease have been made.

## 5.1. Analysis of chlorinated dibenzo-*p*-dioxins and dibenzofurans

Few environmental applications illustrate the capabilities of GC–MS–computer for quantitative trace organic analysis better than the determination of polychlorinated dibenzo-*p*-dioxins (PCDDs) and a related class of compounds, the polychlorinated dibenzofurans (PCDFs). Fig. 5.1 shows the basic structure and number of compounds having from one to eight chlorine substituents. PCDDs or PCDFs having different numbers of chlorines are called congeners, while those with the same number of chlorines are called isomers. There are 75 PCDD congeners, while there are 22 PCDD isomers having four chlorine atoms. Examples of the analysis of

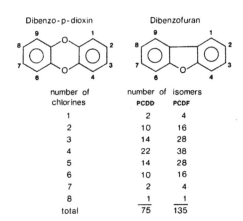

| number of chlorines | number of isomers | |
|---|---|---|
| | PCDD | PCDF |
| 1 | 2 | 4 |
| 2 | 10 | 16 |
| 3 | 14 | 28 |
| 4 | 22 | 38 |
| 5 | 14 | 28 |
| 6 | 10 | 16 |
| 7 | 2 | 4 |
| 8 | 1 | 1 |
| total | 75 | 135 |

Fig. 5.1. Basic structures of PCDD and PCDF.

PCDDs also appear in other chapters, notably in section 4.5, as illustrations of how GC–MS quantitative analysis using selected ion monitoring (SIM) is performed.

These compounds are of interest because of the high toxicity of some of them. The most toxic PCDD/PCDF compound is 2,3,7,8-tetrachlorodibenzo-p-dioxin (2,3,7,8-TCDD), which demonstrated an $LD_{50}$ of 0.6 $\mu$g per kg body weight for the guinea pig. Extrapolation of this toxicity to humans would make 2,3,7,8-TCDD one of the most toxic substances known. Other PCDDs/PCDFs have also demonstrated low $LD_{50}$ values in laboratory animal studies. To protect human health, analytical methods should be able to detect PCDDs/PCDFs compounds in the environment at ppt levels.

### 5.1.1. Incinerator emissions

Incineration of municipal garbage is an important means of waste management. The main products of this process are non-combustible solids and light ash particles (fly ash) that are produced in the combustion zone. About 98% are separated from the effluent stack gases by electrostatic precipitation, then usually disposed in landfills. Incinerators of modern design can also recover significant quantities of useable energy from the combustion. A small fraction of the fly ash particles, as well as vapors which include water, inorganics such as HCl, and volatilized organic compounds, are not collected by pollution abatement equipment and are emitted to the atmosphere through the smoke stack.

Among the trace organics in these emissions are many of the PCDDs and PCDFs. Their analysis is complicated by the presence of as many as 400 other substances, some of which are present at concentrations thousands of times greater. Complex sampling and sample preparation procedures are required to obtain samples for GC–MS analysis that are representative of the emissions from muncipal incinerator stacks. The result of these procedures is to trap a fraction of the emissions over typical sampling periods of 4–24 h and recover the PCDDs/PCDFs by extraction and condensation into final sample volumes of 50–100 $\mu$l. Quantitative analysis by GC–MS is by SIM. By careful choice of GC conditions and use of modern data system techniques, all of the PCDDs and PCDFs containing four or more chlorine atoms can be analyzed in a single sample injection.

*Choice of ions.* Characteristic ions for SIM analysis are chosen by selecting ions from the full mass spectra of representative compounds of each congener group. Fig. 5.2, for example, is the mass spectrum of 2,3,7,8-TCDD. The three most abundant ions are at $m/z$ 320, 322, and 324 of the molecular ion cluster, and these are the ones used for GC–MS–SIM analysis. Generally, quantitation is performed using the most abundant ion, in this case $m/z$ 322, while the others are used for confirmation. As confirming ions, they must be detected at the same retention times and have identical peak profiles as those of the quantitation ion. Ratios of peak areas must be the same as the relative abundances observed in the full mass spectrum. For 2,3,7,8-TCDD, the 320:322:324 peak areas must have the relative ratios 78:100:48, within experimental error ($\pm 10$–15%). Peaks belonging to the molecular ion cluster of a specific PCDD/PCDF group of isomers have area ratios determined by the number of chlorine atoms present, therefore they can be

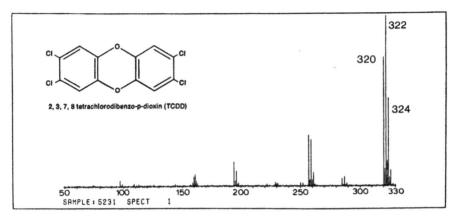

Fig. 5.2. Mass spectrum of 2,3,7,8-TCDD.

calculated even without obtaining mass spectra. Other confirmation ions could be used. For TCDD isomers, for example, the principal decomposition ion is at $m/z$ 257, and is a result of the loss of [COCl] from the molecular ion.

Table 5.1 lists the $m/z$ values chosen for the analysis of PCDDs and PCDFs containing four to eight chlorine atoms, and gives their theoretical ratios. While nominal (integer) masses are used for convenience during description of the analysis method, the exact masses are not integers. Ions have been divided into five groups of six ions each so that high sensitivity can be obtained. The detection limits achievable depend upon the dwell time of each ion and cycle time for each group of ions in a SIM program. Five separate groups of ions are adequate to cover the PCDDs/PCDFs of interest because groups of PCDDs/PCDFs containing different numbers of chlorine atoms are separated from each other by GC.

TABLE 5.1
PCDD-PCDF MASSES MONITORED FOR SIM ANALYSIS
RA = relative abundance. Analysis based on isotopic masses: C = 12.0000, 0 = 15.9949, H = 1.0078, Cl = 34.9689.

| Tetra | | Penta | | Hexa | | Hepta | | Octa | |
|---|---|---|---|---|---|---|---|---|---|
| $m/z$ | RA | $m/z$ | RA | $m/z$ | RA | $m/z$ | RA | $m/z$ | RA |
| Chlorinated dibenzo-p-dioxins | | | | | | | | | |
| 319.9 | 78 | 353.9 | 61 | 387.8 | 51 | 421.8 | 44 | 457.7 | 88 |
| 321.9 | 100 | 355.9 | 100 | 389.9 | 100 | 423.8 | 100 | 459.7 | 100 |
| 331.9 * | 78 | 357.9 | 65 | 391.8 | 81 | 425.8 | 97 | 469.7 * | 88 |
| 333.9 * | 100 | | | | | | | 471.7 * | 100 |
| Chlorinated dibenzofurans | | | | | | | | | |
| 303.9 | 78 | 337.9 | 61 | 371.8 | 51 | 405.8 | 44 | 441.7 | 88 |
| 305.9 | 100 | 339.9 | 100 | 373.8 | 100 | 407.8 | 100 | 443.7 | 100 |
| | | 341.9 | 65 | 375.8 | 81 | 409.8 | 97 | | |

* [13]C-labelled internal standard.

```
MID              DESC: L4
INST: FINN45    CALI: C50506A

MASS DEFECT AT 100 AMU      30 MMU
MASTER RATE               4096
TOTAL ACQU TIME           1.265 SECS
TOTAL SCAN TIME           1.291 SECS
CENT SAMP INT             0.200 MS
CALI MASS RANGE             69 TO   502 AMU
  6.    303.649    334.150      1.000    1.291    1    80    O    1    O    POS
  INT     BEGIN       END        TIME   (SECS)   MPW   MFW   MA   TH   BL   ION
   #       MASS       MASS      REQUEST  ACTUAL

  1.     303.649    304.150      0.200    0.210    5    80    O    1    O    POS
  2.     305.649    306.150      0.200    0.210    5    80    O    1    O    POS
  3.     319.649    320.150      0.200    0.210    5    80    O    1    O    POS
  4.     321.649    322.150      0.200    0.210    5    80    O    1    O    POS
  5.     331.649    332.150      0.200    0.210    5    80    O    1    O    POS
  6.     333.649    334.150      0.200    0.216    5    80    O    1    O    POS
```

Fig. 5.3. Program used for GC–MS–SIM analysis of tetrachlorinated dioxins and furans

*Description of the GC–MS method.* A Finnigan MAT 4500 quadrupole GC–MS system with Incos data system and 30 m WCOT fused silica DB-5 column can be employed to do this analysis. GC conditions are: sample size, $2\mu l$; splitless injection mode; injection temperature, 260°C; initial GC temperature, 80°C for 2.0 min; program rate, 15°C/min to 250°C, then 5°C/min to 300°C and hold for 10 min. The GC–MS transfer line is at 300°C and the mass spectrometer is operated in electron impact mode with electron energy 32 eV. This value was found from previous work to give optimum sensitivity for the analysis of PCDDs/PCDFs on this instrument. A procedure language computer program is used to acquire data and control the switching of groups of ions, or descriptors, at the proper times. This procedure was discussed in section 4.5.1 (procedure PCDD45), and an example of an individual program is shown in Fig. 5.3, which is the program for the tetrachlorinated dibenzo-*p*-dioxins (TCDDs) and tetrachlorinated dibenzofurans (TCDFs).

Fig. 5.3 shows six $m/z$ values of the descriptor L4 program which is created by the program MID of the Incos data system (section 4.5.1). Other key parameters such as dwell times, data acquisition rate, and parameters relating to the mass peak detection threshold are also listed in L4. For each $m/z$ value listed a range is given rather than the exact mass. This range of 0.5 a.m.u. is centred around the exact mass, and ensures that the top of each mass peak will not be missed. Although the total time requested for one cycle of the L4 program is 1.2 s (200 ms per ion), the actual total scan time required is 1.3 s.

*Results and discussion.* Fig. 5.4 is a total ion plot of all 30 ions monitored during the analysis of a 2.0-$\mu l$ injection of a cleaned-up toluene extract of the particulate matter emitted from the stack of a municipal incinerator. The elution times covered by each group of ions are indicated on the plot. Each of the five regions contains the scans made using an individual program containing six $m/z$ values such as shown in Fig. 5.3. To create the total ion plot, the total ion abundances for individual scans were summed, in the same manner as the reconstructed gas chromatogram (RGC) is

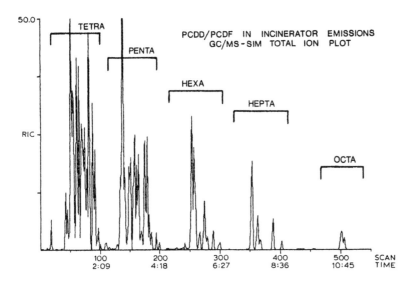

Fig. 5.4. Total ion plot from GC–MS–SIM analysis of incinerator emissions for PCDDs and PCDFs.

plotted using stored mass spectra. A total of 72 peaks are detected, all of which are PCDDs or PCDFs. Because only two or three characteristic $m/z$ values for each PCDDs or PCDFs are monitored, strict criteria are used to ensure peaks are not due to interferences. These criteria are as follows:

(1) coincident response for two or three ions monitored,

(2) peak falls within known elution time window,

(3) ratios of coincident peak areas given for two or three ions monitored are within 10% of theoretical values given in Table 5.1,

(4) all peaks are from compounds that have been through cleanup procedures designed to remove interferences,

(5) signal-to-noise of peaks must be better than 3:1.

These criteria are illustrated by Fig. 5.5, which is a plot of the three ions used for the pentachlorinated dibenzo-$p$-dioxins ($P_5$CDDs). The patterns of each $m/z$ value plotted are identical. Area ratios of corresponding peaks in the three plots are 63:100:61, which are well within experimental variation from the theoretical values of 61:100:65. Each peak in Fig. 5.5 represents a $P_5$CDD isomer, and at least 11 of the possible 14 $P_5$CDD isomers are present, although the chromatographic peaks are not fully resolved.

Quantitation is performed by comparing peak areas with areas of external standards. Correction for recoveries is made using the [$^{13}$C]-2,3,7,8-TCDD and [$^{13}$C]-OCDD internal standards which are spiked into the sample before extraction. Both spikes are used for this correction because the higher chlorinated species do not always behave in exactly the same manner as the lower chlorinated congeners. Ideally, an internal standard representative of each congener group should be employed. Quantitative results of this analysis are shown in Table 5.2. Although the

166

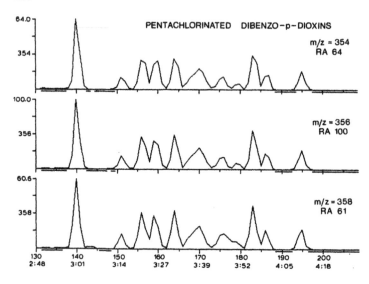

Fig. 5.5. Plot of three characteristic ions of pentachlorinated dioxin isomers.

totals for each group of isomers varied from 2–30 ng, the groups, except for octachlorinated, are all composed of 2–15 isomers. Therefore, many of the individual peaks shown in Figs. 5.4 and 5.5 represent less than 1.0 ng injected. Even at these low levels, none of the peaks are close to the detection limits achievable. Using the identical instrumentation described here, limits of detection for 2,3,7,8-TCDD of 1.0 pg injected are possible. On a daily basis, without special optimization of all parameters including source cleaning, detection limits for 2,3,7,8-TCDD vary from 5–10 pg injected.

TABLE 5.2
PCDDs AND PCDFs IN STACK EMITTED PARTICULATES BY GC–MS–SIM

| Congener | PCDD | | PCDF | |
|---|---|---|---|---|
| | Number of isomers | Amount detected (ng) | Number of isomers | Amount detected (ng) |
| Tetra | 13 | 5.3 | 15 | 33 |
| Penta | 12 | 6.7 | 14 | 33 |
| Hexa | 6 | 7.3 | 6 | 17 |
| Hepta | 2 | 6.8 | 2 | 11 |
| Octa | 1 | 3.6 | 1 | 1.8 |

*5.1.2. Biological samples: fish*

Biological matrices are among the most difficult to analyze for trace levels of organic compounds, however, tissue from fish and bird species is an important indicator of environmental contamination. Often, trace contaminants are found in

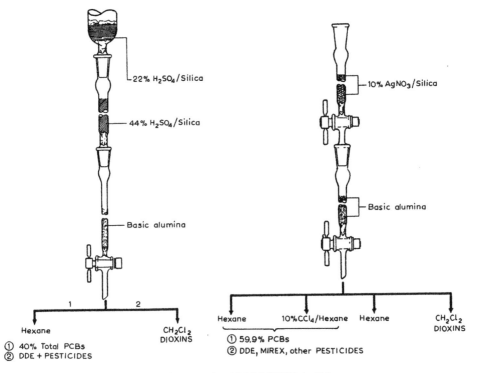

Fig. 5.6. Column cleanup procedure for analysis of 2,3,7,8-TCDD in fish.

such samples through bioconcentration even when direct monitoring of air, water, or soil fails to detect such contaminants. Fish are especially important since they are directly consumed by humans. Fish do not have the range of PCDDs and PCDFs found in incinerator emissions, and most monitoring is performed only for 2,3,7,8-TCDD isomer-specific determinations. Because the 22 TCDD isomers all have virtually identical mass spectra, special procedures are required to assure that the 2,3,7,8-TCDD isomer is the one detected and quantified.

*Sample extraction and cleanup.* Before extraction, fish tissue is spiked with $[^{13}C]$-2,3,7,8-TCDD and the bulk of the tissue is destroyed by concentrated acid digestion. The 2,3,7,8-TCDD is removed from aqueous solution by solvent extraction and most co-extractives are removed by performing column chromatography cleanup. Fig. 5.6 illustrates the cleanup procedure used, which is similar to that used for incinerator samples. The acid silica destroys most compounds that are co-extracted with 2,3,7,8-TCDD. Silver nitrate traps sulfur-containing compounds and alumina separates 2,3,7,8-TCDD from other chlorinated organics such as polychlorinated biphenyls. In addition, high performance liquid chromatography (HPLC) is used to isolate 2,3,7,8-TCDD from other compounds that are still present. Fig. 5.7 shows how HPLC cleanup is performed. A specific portion of the eluate known to contain 2,3,7,8-TCDD is collected, and the remaining eluate can be collected and

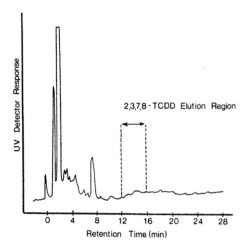

Fig. 5.7. HPLC cleanup of fish extract for 2,3,7,8-TCDD analysis.

analyzed for other compounds or discarded. The HPLC elution time window is established by analysis of standards. A narrow window gives better cleanup but if too narrow there is increased risk of removing some of the 2,3,7,8-TCDD or missing it entirely. No peaks are observed in the indicated elution windows of Fig. 5.7 because the concentrations of 2,3,7,8-TCDD and [$^{13}$C]-2,3,7,8-TCDD, are too low to give an observable response on the UV detector employed.

*Results and discussion.* The GC–MS instrumentation and conditions for analysis of 2,3,7,8-TCDD in the collected HPLC fraction are the same as described previously, except only one descriptor containing six *m/z* values is needed. Fig. 5.8

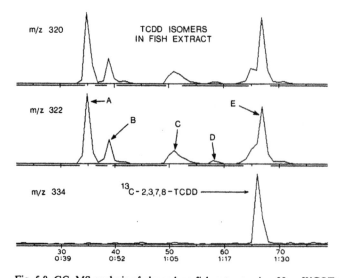

Fig. 5.8. GC–MS analysis of cleaned-up fish extract using 30-m WCOT GC column.

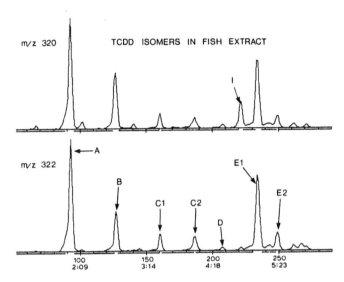

Fig. 5.9. GC–MS analysis of cleaned-up fish extract using a 60-m WCOT column that resolves 2,3,7,8-TCDD from other TCDD isomers.

shows the GC–MS analysis of the HPLC cleaned-up sample. The $m/z$ 334 internal standard coincides with that of peak E in Fig. 5.8. The analysis still cannot be considered to be isomer-specific for 2,3,7,8-TCDD, however, because several other TCDD isomers are also present, and the 30-m column employed cannot resolve 2,3,7,8-TCDD from some of these other isomers. It can be seen that peak E is not a single peak, it contains other unresolved TCDD isomers and interferences.

By using a 60-m WCOT column that can resolve 2,3,7,8-TCDD from the other 21 isomers, isomer-specific determination of this compound is obtained. Fig. 5.9 shows the analysis of the same sample on such a column. Peaks corresponding to those in Fig. 5.8 are indicated. Peak C in Fig. 5.8 is composed of two TCDD isomers (C1,C2), while peak E contains 2,3,7,8-TCDD (E1) as well as a second TCDD isomer (E2) and an interference in the $m/z$ 320 ion (peak I). Quantitation of 2,3,7,8-TCDD can be performed using the internal standard only, since a known amount of [$^{13}$C]-2,3,7,8-TCDD was added to the sample before extraction. Correction for percent recovery through the sample cleanup is then automatically taken into account. The elution times of [$^{13}$C]-2,3,7,8-TCDD and native 2,3,7,8-TCDD are not identical. The difference is only a few seconds, however, and does not affect the ability to correctly identify the 2,3,7,8-TCDD isomer if the 60-m column is used and if both native and $^{13}$C-labelled standards are analyzed individually to determine the exact retention time difference.

### 5.1.3. Summary and conclusions

The applications presented here illustrate the power of GC–MS to perform quantitative analysis of trace levels of organic compounds in complex mixtures. Comparison of the data in Figs. 5.7–5.9, show that GC–MS detected several peaks at very good signal to noise ratios in a sample that produced only background response in a UV detector. Analysis of incinerator emissions shows the importance of the data system, which permitted the ion switching necessary to analyze many compounds by SIM in a single sample injection. The analysis of fish tissue illustrates the importance of chromatography in isomer-specific determinations.

An important consideration is the sample treatment prior to GC–MS analysis. The ability of GC–MS to provide quality control checks on analytical procedures through the use of isotopically labelled analogues of the analytes is a powerful technique. Such quality control procedures insure that a high degree of confidence can be placed in the quantitative data obtained.

## 5.2. Use of high performance liquid chromatography for complex mixture analysis

In the previous section it was shown how high performance liquid chromatography (HPLC) is used for the isomer-specific analysis of 2,3,7,8-TCDD in fish tissue. HPLC can perform separations of complex organic mixtures that are not possible with GC. Its applicability to complex sample analysis is much broader in scope than presented previously. By separating mixtures into distinct fractions containing specific compound classes, such as hydrocarbons, aromatics, phthalate esters and chlorinated aromatics, the complexity and interferences in the individual fractions are much reduced from the original mixture. HPLC instrumentation is now so well developed and computerized that preseparations of these types are quite feasible. In the previous section the analysis of PCDDs/PCDFs in stack-emitted particulates was described. The analysis of these and other compounds in the particles collected by pollution equipment before reaching the stack will be shown here to illustrate in detail the use of HPLC to simplify GC–MS analysis.

### 5.2.1. HPLC fractionation of the mixture into compound types

A one-step separation procedure using normal phase HPLC with gradient elution can be employed for the analysis of PCDDs in a complex mixture. The feasibility of this method is demonstrated by the analysis of PCDDs in a fly ash sample from a municipal incinerator in which more than 300 organic components are present as indicated by the number of peaks in a gas chromatogram. A 200-g sample of fly ash is extracted with benzene in a Soxhlet extractor for 48 h, and the concentration of extract is adjusted to be equivalent to ca. 0.3 g fly ash per $\mu$l for HPLC separation.

The instrument used for separation is a Spectra-Physics SP-8000 HPLC equipped with SP-8400 UV/Vis detector and SP-4100 integrator. The monitoring wavelength is 254 nm. A 10 $\mu$m semi-preparative Spherisorb silica column and 140 $\mu$l sample loop are used. A gradient elution program using n-$C_6H_{14}$, $CH_2Cl_2$ and $CH_3CN$ was developed to separate the sample of fly ash extract, using a carrier solvent flow rate of 5 ml/min.

Fig. 5.10 is a diagram showing the gradient program used (upper trace), and the fraction collection interval and HPLC profile of fly ash extract (lower trace). Each fraction is separately collected and then subjected to detailed analysis by GC–MS techniques.

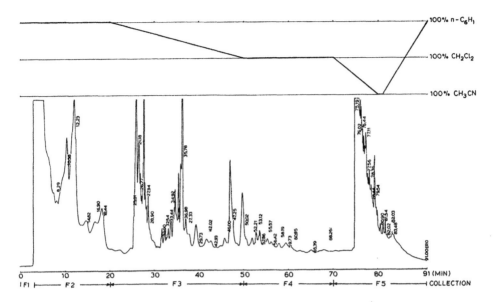

Fig. 5.10. HPLC fractionation of fly ash extract.

*Qualitative analysis of HPLC fractions.* The qualitative analysis of each fraction was performed using a Hewlett-Packard 5992 GC–MS system equipped with a 30 m X 0.32 mm I.D. DB-5 fused silica WCOT column. Compound identification was done by matching sample mass spectra obtained from the sample to standard spectra from printed collections of reference spectra, or from computer library searches.

Retention indices based on polycyclic aromatic hydrocarbon reference compounds have also been used to facilitate compound identification. Upon completion of a sample GC run the retention indices of components in fractions were automatically calculated on a GC equipped with the same WCOT column used for GC–MS analysis. Over 200 organic components were identified in the five HPLC fractions collected.

*PCDDs eluted exclusively in fraction 2.* The normal phase HPLC and gradient program employed separates the organic compounds in the fly ash extract according to their relative polarities. Table 5.3 lists the principal compound classes identified in each fraction.

The second fraction is important because it contains a number of components having high environmental significance. Fig. 5.11 is the WCOT GC trace of HPLC fraction 2 of the fly ash extract. This trace was duplicated by the total ion current trace obtained from GC–MS analysis of fraction 2 using the same column and chromatographic conditions. Ninety-two components were identified in this fraction and their retention indices were calculated with an average standard deviation of 0.023 based on triplicate GC injections. Table 5.4 is a listing of the components identified in fraction 2. The numbers in Table 5.4 refer to those seen on the GC trace in Fig. 5.11.

TABLE 5.3

PRINCIPAL COMPOUND CLASSES IDENTIFIED IN HPLC FRACTIONS OF FLY ASH EX-
TRACT

| Fraction number | Type of compound |
|---|---|
| 1 | Normal and branched chain hydrocarbons |
| 2 | Polyaromatic hydrocarbons (PAHs) with two or three rings |
| | Sulfur-PAHs |
| | Polychlorinated PAHs with one-three rings |
| | PCDDs |
| | PCDFs |
| 3 | Oxy-PAHs |
| | Nitro-PAHs |
| | Polychlorinated oxy-PAHs |
| | PAHs with three or more rings |
| 4 | Oxy-PAH |
| | Phthalates |
| 5 | Benzo[c]cinnoline |
| | High-molecular-weight phthalates |

   PCDDs exclusively elute in fraction 2 under these HPLC conditions. This was
confirmed by analysis of all the fractions collected from the HPLC runs of a
mixture of PCDD standards and also of the fly ash extract using GC–MS–SIM. No
PCDDs were detected in other fractions when overloading of the HPLC column was
avoided.
   *Quantitation of PCDDs.* Fraction 2 can be used for the quantitation of PCDDs in
the fly ash extract by GC–MS–SIM analysis using external standards. This was

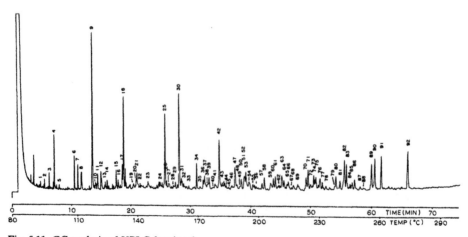

Fig. 5.11. GC analysis of HPLC fraction 2.

TABLE 5.4
TYPES OF COMPOUNDS IDENTIFIED IN HPLC FRACTION 2 OF FLY ASH EXTRACT

| GC peak number | Compound type |
|---|---|
| 1,2,9,15,20,24<br>4–7,10,11,12–14<br>19,26,30,34,36,38<br>39,42,44,47,68 | Alkyl substituted benzene and biphenyl,<br>PAHs and alkyl-substituted PAHs |
| 3,8,18,21,23,25<br>27,29,31,33,37,40<br>43,50,52,54,56,57<br>64,87,88 | Polychlorinated PAHs |
| 17 | Dibenzofuran |
| 28 | Sulfur-PAHs |
| 35,45,46,48,49,53<br>58,59,62,69,70,72,79<br>80,84,89 | PCDFs |
| 41,51,55,60,61,63<br>65–67,71,73<br>74–78,81–83,85,86<br>90–92 | PCDDs |

done by comparing total areas of PCDD peaks to areas of individual PCDD standards representative of each chlorinated group. These quantitative results for PCDDs in the fly ash are listed in Table 5.5. The average recovery of the PCDD standard mixture through HPLC fractionation was 105%.

These results indicate that this one-step HPLC separation for analysis of PCDDs in complex mixtures, such as the fly ash extract, is very effective. Compared with conventional column chromatography clean-up methods, HPLC has the advantages of speed, simplicity, and high efficiency. Further, autosamplers and automatic fraction collectors are available so that the entire procedure can be automated.

Additional separation of components in fraction 2 can be achieved using re-versed-phase HPLC fractionation. This is shown in Fig. 5.12. Using this second

TABLE 5.5
CONCENTRATION OF PCDD CONGENER GROUPS IN FLY ASH

| Congener group | Number of isomers | Concentration (ppb) |
|---|---|---|
| Tetrachlorinated | 22 | 540 |
| Pentachlorinated | 14 | 470 |
| Hexachlorinated | 10 | 590 |
| Heptachlorinated | 2 | 430 |
| Octachlorodibenzo-p-dioxin | 1 | 470 |

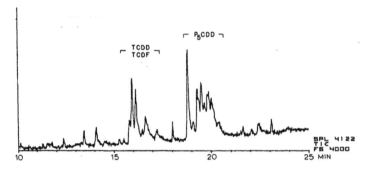

Fig. 5.12. GC–MS–SIM analysis of fraction 2 after additional reversed-phase HPLC separation.

step, the tetra- and penta-PCDD congeners are completely separated from the original mixture of over 300 compounds.

## 5.3. Use of high resolution GC–MS for identity confirmation

Most GC–MS determinations of trace level target compounds in complex samples are performed using quadrupole systems. This is because of their lower initial cost, ease of use, and generally greater sample throughput, compared to magnetic systems. Since only a few selected ions are chosen for GC–MS–SIM analysis, however, interfering ions having the same nominal (integer) masses as analyte ions can cause problems in the identification of peaks.

Such interferences can cause the false rejection of analytes by making ion ratios incorrect in the case where multiple characteristics ions are monitored (false negative). The opposite problem of identifying an interference as the analyte (false positive) can also occur, especially if only one characteristic ion is monitored. In such cases it may be possible to eliminate the interference by high resolution mass spectrometry (HRMS).

Fig. 5.13 is a plot of ion abundances of two masses characteristic of TCDD isomers from the GC–MS–SIM analysis of a concentrated solvent extract of a soil sample. A low resolution quadrupole analyzer was used for this work. The GC–MS methods employed were described in the previous section. Twenty-two peaks are observed that have the correct theoretical ratios of 78:100 for the $m/z$ 320:322 ions characteristic of TCDD. These peaks are numbered in the figure.

The results were suspect because it was known from previous work that the 30-m WCOT column used cannot separate all 22 TCDD isomers to the degree shown in Fig. 5.13. To confirm the finding obtained by the low resolution quadrupole mass spectrometry (LRMS), a high resolution, double-focussing VG ZAB–HS mass spectrometer operated at a resolution of 12,000 and VG 11/250 data system were used for this analysis. A Varian 6000 GC system equipped with on-column injector, and the same type of WCOT column used for the LRMS analysis were employed. The GC column was directly coupled to the MS ion source.

Characteristic ions of TCDDs were monitored at $m/z$ 319.8965 and 321.8936.

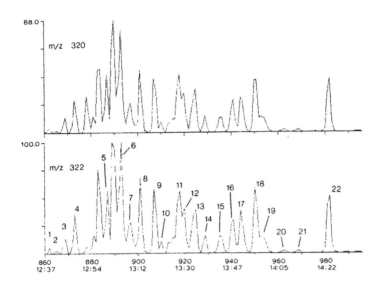

Fig. 5.13. GC–MS–SIM analysis of TCDD isomers in soil sample extract.

Ions of $m/z$ 331.9368 and 333.9338 characteristic of the [$^{13}$C]-TCDD internal standard were also monitored. Ion dwell times were 50 ms and a 15-ms delay time was required between each jump in mass to stabilize the system. For medium resolution determination, a lock mass of 300.0939 from coronene was used. The coronene was bled into the ion source throughout the analysis, and mass measurements were made by accurate comparison of peaks to the coronene lock mass.

Fig. 5.14 shows the analysis of the same sample as Fig. 5.13 but using the VG-ZAB operating at a resolution of 12,000. No peaks are observed in this analysis, only background noise. For this determination, the limit of detection for the soil

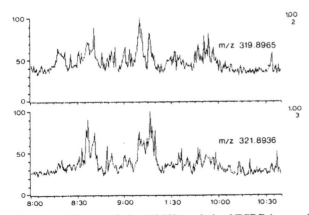

Fig. 5.14. Medium resolution (12,000) analysis of TCDD isomers in soil sample extracts.

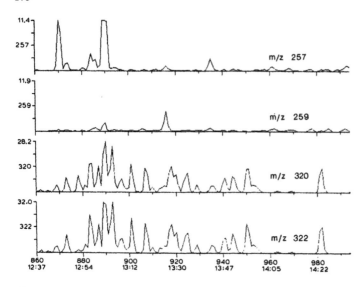

Fig. 5.15. LRMS analysis of TCDD isomers in soil sample extract using four characteristic ions.

sample analyzed is about 0.5 ppt (0.5 pg/g). From the LRMS determination, a concentration of total TCDD isomers is calculated to be 20 ppb (20 ng/g). A difference of a factor of 40,000 between these data resulted from interferences in the LRMS determination that were eliminated by the HRMS system.

Another approach to the problem of interferences using only LRMS is to include additional characteristic ions in the GC–MS–SIM analysis. For TCDD isomers, a characteristic mass fragment is formed due to the loss of [COCl] from the parent ion. Peaks at $m/z$ 257 and 259 are observed for this fragment ion, at relative abundances of about 30% of the base peak ($m/z$ 322). Fig. 5.15 shows the re-analysis of the same sample as Fig. 5.13 and 5.14 but using four characteristic TCDD ions. The $m/z$ 320:322 patterns are identical to those observed in Fig. 5.13, however, none of the observed peaks corresponds to peaks in both $m/z$ 257 and 259 plots. They have been normalized so that TCDD isomers will give an observed peak the same shape and size as observed in the $m/z$ 320:322 plots. Since no such peaks are observed, it is concluded that no TCDD isomers are present in this sample extract.

LRMS is adequate for most target compound analysis at trace levels, but analytical protocols should include routine checks on the accuracy of identifications based only on a few ions. These protocols can include submitting some samples to HRMS analysis for confirmation by operating at a high enough mass resolution so that known interferences will not be detected. Used in this manner, the techniques of LRMS and HRMS are complimentary, and both the high throughput of the LRMS and special capabilities of the HRMS can be used to maximize the quantity and quality of GC–MS data.

## 5.4. New developments: MS–MS

Although GC–MS is in an advanced stage of development, new techniques, applications, and instrumentation are continually being introduced. Many of these are minor modifications of existing methods, or are for specialized uses such as extended mass range and the analysis of substances of low volatility. New developments in such diverse fields as gas chromatography, ionization methods, mass analysis, computer techniques, and mathematical procedures such as pattern recognition are all of potential value to GC–MS. Among the most significant recent developments are the techniques of tandem mass spectrometry (MS–MS) and Fourier transform mass spectrometry (FTMS). These developments show how unique application of existing technology can lead to new analytical capabilities. Computers are essential to both MS–MS and FTMS because of the vast amounts of data that have to be processed — even more than for conventional GC–MS analysis.

The power of GC–MS arises from the very high sensitivity of the mass spectrometer and selectivity from two types of separations. First, compounds in the vapour state are separated by a chromatographic column on the basis of their chemical and physical properties. In the mass spectrometer ion source, characteristic ions of each substance are formed. These ions then are separated from each other and detected according to their $m/z$ ratios. Because of the two types of separation (molecules and ions) the GC–MS instrument is both a universal and tunable highly specific detector. Selectivity can be increased by greater GC resolution or greater MS resolution.

Initial chromatographic separation is slow compared to the ion separation in the mass spectrometer. For a complex sample it is not unusual to take an hour or more for a GC run, while the complete mass spectrum of a compound can be generated in less than 1 s. Also, in spite of the good resolving power of WCOT columns, unambiguous determination of specific substances may still require pre-separation of interferences using HPLC or column chromatography before GC–MS determination. Such cleanup steps may take days to complete.

A large savings in time can be obtained by replacing the gas chromatograph with a second mass spectrometer. This new technique, called tandem mass spectrometry or MS–MS, still retains the advantages of two separations of sample components. In MS–MS, however, both separations are of ions. Because the specificity of MS–MS is higher than for GC–MS, quantitative analysis of known analytes in simple mixtures can be performed in a manner of minutes without chromatographic separation or other chemical treatment to remove interferences. In addition to this direct mixture analysis, MS–MS can be applied to a wider range of samples than GC–MS since analytes do not have to be volatile enough to pass through a GC column.

### 5.4.1. Description of the technique

Fig. 5.16 illustrates how the technique is practiced. A four component mixture consisting of the hypothetical molecules ABC, BCA, ABZ, and XYZ is introduced

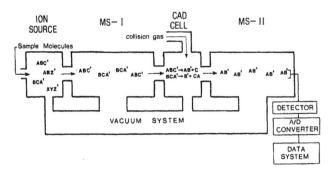

Fig. 5.16. Principle of MS–MS.

into the ion source where molecular ions ABC$^+$, BCA$^+$, ABZ$^+$ and XYZ$^+$ are formed. If the first mass filter (MS-I) is set to pass only ions of $m/z$ = ABC$^+$, then only ABC$^+$ and BCA$^+$ will be selected. These ions are allowed to enter a reaction region where they are made to fragment into other ions by collision with a neutral target gas. This process is called collision-activated dissociation (CAD). Since ABC$^+$ and BCA$^+$ have different structures, they will decompose in the CAD cell to form different species, for example:

$$ABC^+ \overset{N}{\to} AB^+ + C$$

$$BCA^+ \overset{N}{\to} B^+ + CA$$

where N indicates decomposition has resulted from collision with a neutral gas molecule. Now if these ions are accelerated into a second mass filter (MS-II) that is set only to pass ions of $m/z$ = AB$^+$ to a detector, then the signal measured will be indicative of the number of molecules of ABC in the original mixture. Therefore, an instrument consisting of (ion source) (MS-I) (CAD cell)  (MS-II) is capable of analyzing specific components of mixtures without prior chromatographic separation.

The preceding example represents a very simple case since many more ions will be present initially, especially if an electron impact ion source is employed. Any ion source used for GC–MS can be employed in MS–MS, however, chemical ionization or other soft ionization techniques are generally preferred because of their reduced initial fragmentation. MS–MS can be viewed as taking a mass spectrum of one ion in a mass spectrum. The quantity of data that can be obtained from even a single compound is immense and requires advanced computer techniques for the analysis. Because MS-I and MS-II can be independently operated in either scanning or selected ion modes, certain types of data can be obtained in MS–MS that are not possible with conventional GC–MS.

*5.4.2. Scan techniques*

In the above example, both MS-I and MS-II were operated in the selected ion monitoring mode. First, MS-I choose the $m/z$ value of ions allowed to pass into the CAD cell. These ions, called parent ions, were induced by collision with neutral gas molecules to decompose into daughter ions. MS-II then choose the $m/z$ value of daughter ions to be detected. This technique, called single reaction monitoring, would be used when a known substance is to be detected in a mixture. It is analogous to SIM in GC–MS but by replacing the GC with a second MS analyzer, an analysis can be performed much more rapidly. Careful choice of ions based on previous experiments using pure substances is necessary. While chemical interferences are fewer in MS–MS than GC–MS, such interferences can still be present.

Other types of data are available. By operating MS-II in the scanning mode and MS-I in the SIM mode, a spectrum is obtained of all daughter ions arising from a specific parent. Such daughter ion spectra are analogous to taking mass spectra of selected ions in a mass spectrum, and are used primarily to identify individual compounds in a complex mixture.

Parent ion spectra are obtained by scanning MS-I but operating MS-II in SIM mode. Parent ion spectra give the parent ions which fragment in the CAD cell to form a specified daughter ion. This technique is used to screen a sample for a specific class of compounds, if members of the compound class are known to form a characteristic ion.

Another method of screening mixtures for specific compound classes is by neutral loss spectra. In this technique, both MS-I and MS-II are scanned with a constant mass difference. This mass difference is chosen to be the mass of a neutral species lost in the CAD cell that is characteristic of a class of compounds or specific functional group. For example, to detect the neutral loss of methyl groups, if MS-I was scanned from $m/z$ 30–500, then MS-II would be scanned from $m/z$ 15–485. The scans of both analyzers would be linked so that the mass difference was always 15.

Table 5.6 is a summary of the scan techniques available with MS–MS. It is also possible to perform the same types of data collection as used in conventional

TABLE 5.6

MS–MS SCAN TECHNIQUES

| Spectrum type | Analyzer mode of operation | | Principal application |
|---|---|---|---|
| | MS-I | MS-II | |
| Single reaction monitoring | SIM | SIM | Analysis of specific compound |
| Daughter ion | SIM | Scanning | Compound identification |
| Parent ion | Scanning | SIM | Compound class analysis |
| Neutral loss | Scanning (mass M1 → M2) | Scanning (mass M1 − x → M2 − x) | Compound class analysis function group analysis |

GC–MS. Because of the many different experiments that can be performed using MS–MS, the abundance of data generated is great. Choice of which scan method to use depends upon the type of information needed and the specific compounds in the sample matrix.

### 5.4.3. Instrumental configurations

MS-I and MS-II do not need to be the same type of analyzer. Almost any combination of MS analyzers can be employed, with the exception of consecutive electrostatic analyzers. Most research has been conducted using combinations of quadrupole (Q), magnetic (B) and electrostatic (E) analyzers to give such MS-I–MS-II configurations as Q–Q, Q–B, B–Q, B–E and E–B. It is usually preferable to have a magnetic or quadrupole analyzer first. A Q–Q–Q (triple quadrupole) configuration has also been used in which the middle quadrupole performs the function of the CAD cell, and is not used as a mass filter.

The specific configuration used affects the performance of an experiment since the quadrupole filter acts on the basis of mass-to-charge ($m/z$) ratio, the magnetic analyzer according to the momentum-to-charge ($mv/z$) ratio, and the electrostatic analyzer by the kinetic energy-to-charge ($mv^2/z$) ratio. Therefore, to perform a specified MS–MS experiment, MS-I and MS-II cannot be scanned in the same manner for all instrumental configurations.

### 5.4.4. Comparison of GC–MS and MS–MS

All possible applications of MS–MS have not been explored. Since all of the regular GC–MS techniques such as positive and negative chemical ionization can be used for MS–MS, the number of experiments that can be applied to the analysis of even a single substance is enormous. A partial list of the principal experimental variables is as follows:

(1) analyzer configuration (Q–Q, Q–Q–Q, Q–B, B–Q, B–E, E–B),
(2) scan technique (parent, daughter, neutral loss, regular GC–MS),
(3) ionization method (EI, positive or negative CIMS),
(4) collision cell parameters (choice of collision gas, other cell parameters).

To obtain increased specificity, a gas chromatograph can be added to the MS–MS configuration. With three types of compound selection, an identification by GC–MS–MS can be considered to be definitive.

Table 5.7 is a comparison of the principal features of MS–MS and conventional GC–MS. Major advantages of MS–MS are speed of analysis and increased applicability. A consequence of increased speed of analysis is the ability to determine full calibration curves for quantitative analysis. In regular GC–MS, sample throughput is much lower and many quantitative analyses are based upon one or two-point calibrations. MS–MS applicability to a wider range of analytical problems is due to the different types of data that can be obtained and greater flexibility of sample types to be analyzed. GC–MS is limited to compounds amenable to GC analysis. Of course, the advantages of GC can be used by providing an extra separation technique in the GC–MS–MS instrument.

Fig. 5.17 shows a commercially available GC–MS–MS system. The actual

TABLE 5.7

COMPARISON OF GC–MS AND MS–MS CHARACTERISTICS

| Feature | GC–MS | MS–MS |
|---|---|---|
| Speed of analysis | Slow | Rapid |
| Specificity | High (SIM) | Very high |
| Types of data | Scanning, SIM | Scanning, SIM, parent, daughter, neutral loss |
| Applicability | Compounds must be amenable to GC analysis | Very broad |
| Identification of unknown compounds in mixtures | Good | Good (if GC–MS–MS) Poor (if only MS–MS) |
| Quantitative analysis | Good (SIM) | Very good |
| Sensitivity | High | Comparable to GC–MS |
| Availability of reference data collections | Good | Poor |
| Level of training to maintain and operate system | Medium to very high | Very high |
| Level of training for data interpretation | Medium to high | Very high |

MS–MS analyzer is a triple quadrupole configuration, where the middle quadrupole acts as a CAD cell. The analyzer itself is a small part of the entire instrument, in terms of physical space. This instrument is very similar to the single quadrupole GC–MS instrument manufactured by the same company. In fact, the GC–MS system can be upgraded to the GC–MS–MS configuration by addition of an extra quadrupole, CAD cell, electronics, and MS–MS software.

Fig. 5.17. GC–MS–MS instrumentation system.

GC–MS is still the technique of choice for identification of unknown compounds in mixtures. Since comparison of unknown and reference spectra is required, the many variables and configurations of MS–MS are not necessarily advantages. Spectra must be obtained under the same conditions for comparison, and large MS–MS reference files do not yet exist. Therefore, interpretation of MS–MS data requires a greater level of expertise than does GC–MS data. Operation and maintenance of MS–MS systems also requires additional training. Some of these problems will undoubtedly be alleviated with the development of more sophisticated data systems and MS–MS data reference files, but this development is slow. For the present, MS–MS is best regarded as a specialized form of mass spectrometry that is presently suited for certain applications not amenable to GC–MS or difficult to solve by GC–MS alone.

## 5.5. Suggested reading

1 F.W. Karasek, *Ind. Res./Develop.*, April (1978).
2 E.O. Oswald, P.W. Albro and J.D. McKinney, *J. Chromatogr.*, 98 (1974) 363–448.
3 G.A. Eiceman, R.E. Clement and F.W. Karasek, *Anal. Chem.*, 51 (1979) 2343–2350.
4 H.Hau and K. Biemann, *Anal. Chem.*, 46 (1974) 426–434.
5 S.C. Gates, M.J. Smisko, C.L. Ashendel, N.D. Young, J.F. Holland and C.C. Sweeley, *Anal. Chem.*, 50 (1978) 433–441.
6 F.W. McLafferty, *Acc. Chem. Res.*, 13, (1980), 33–39.
7 F.W. McLafferty, *Science* 214 (1981), 280–287.
8 R.G. Cooks and G.L. Glish, *Chem. Eng. News*, Nov. 30, 1981, 40–52.
9 H.Y. Tong, D.L. Shore and F.W. Karasek, *Anal. Chem.*, 56 (1984) 2442.

# GAS CHROMATOGRAPHY–MASS SPECTROMETRY GLOSSARY

**ABSOLUTE CALIBRATION** - Syn: DIRECT CALIBRATION or EXTERNAL STANDARD - Method relating detector response to sample concentration in order to perform quantitative analysis. Standard solutions of a sample to be quantitated are prepared and equal volumes chromatographed. Peak heights or peak areas are plotted versus concentrations to produce a calibration curve.

**ABSOLUTE RETENTION TIME** - The time elapsed between the introduction of the sample and the appearance of the GC peak maximum.

**ACOUSTIC COUPLER** - Modem that converts data to a sequence of tones for transmission over telephone lines. Used for communication with computers in distant locations.

**ADSORPTION** - Process which occurs at the surface of a liquid or solid as a result of the attractive forces between the adsorbent and solute; the basis for separation in GSC and LSC.

**ADSORPTION - GC–MS SYSTEM** - Certain compounds, which are highly adsorptive and low in concentrations, will adsorb irreversibly on active sites anywhere in the GC–MS system - sample injection components, GC column, GC–MS interface, and MS ion source.

**ALKALI FLAME IONIZATION DETECTOR (AFID)** - Syn: NITROGEN-PHOS-PHORUS DETECTOR (NPD), THERMIONIC EMISSION DETECTOR (TED) - Detector operating by thermal heating of an alkali glass bead source. Very selective response to compounds containing phosphorus or nitrogen is obtained.

**AMPLIFIER** - Syn: ELECTROMETER - Combination of electronic circuits used to amplify and filter the chromatographic signal from ionization detectors.

**ANALOG-TO-DIGITAL (A/D) CONVERTER** - A device used to convert electrical signals (analog) into discrete, representative numbers (digital).

**APIEZON(S)** - High-molecular-weight hydrocarbon grease commonly used as stationary phase in gas chromatography. Non-polar. Has been replaced mostly by methylsilicones.

**ATLAS OF MASS SPECTRA** - Book containing reference mass spectra tabulated by the ionic mass and relative abundance of the eight or ten most abundant ions in the spectrum.

**BACKGROUND SUBTRACTION** - Computerized procedure in which the background ion intensities from mass spectra (column bleed, unresolved minor components) are subtracted from intensities of ions of the sample component to give a net mass spectrum that contains only sample component ions.

**BAND** - Mobile phase zone which contains a sample component.

**BAND WIDTH** - Dispersion of the chromatographic signal, usually measured at the peak half height or two thirds height. Used to calculate the number of theoretical plates.

**BAR GRAPH PRESENTATION** - Data presentation technique in which mass spectra are plotted as a series of vertical lines vs. ionic mass, the heights of lines indicating relative ionic abundances.

**BASELINE** - Constant signal produced by the background level of the instrument. Usually represented by a flat line on the recorder.

**BLEEDING** - Column bleed. Loss of the stationary phase due to its own volatility. Every phase has a maximum operating temperature above which vaporization is excessive. In GC–MS this bleed results in a large background of ions that is inseparable from sample ions.

**BOOTSTRAP ROUTINE** - A special procedure that loads the first few instructions of a program (generally the operating system) into computer storage. These instructions then bring in the remainder of the program. A bootstrap routine is needed to start up the computer system.

**CAPACITY FACTOR** - Syn: CAPACITY RATIO - Ratio of the time the sample spends in the stationary and mobile phases $k = t(r)/t(0)$.

**CAPILLARY COLUMN** - see OPEN TUBULAR COLUMN

**CARBOWAX (CES)** - Polyethylene glycol polymers used as stationary phases in GC. The polar phase Carbowax 20 M is a popular liquid phase.

**CARRIER GAS** - Gas used as the mobile phase in GC.

**CATHODE RAY TUBE (CRT)** - Form of data display for GC–MS systems. Consists of a television-like tube and screen upon which GC–MS data can be displayed in real-time.

**CELITE** - Diatomaceous earth solid support, also used as filtration aid and in making short columns.

**CHEMICAL IONIZATION (CI)** - Method of ionizing sample molecules involving an ion-molecule reaction. The reactant ions are formed by electron ionization of a reagent gas (i.e. methane at a pressure near 1 Torr in the ion source of the mass spectrometer). Both positive and negative ions can be formed.

**CHROMATOGRAM** - Graphical representation of a separation obtained from a chromatograph. The result of plotting detector response vs. time. Can be used for qualitative and quantitative information.

**CHROMOSORB(S)** - Tradename of Johns Manville Corp. given to a family of solid supports obtained mainly from diatomaceous earth. Series includes chromosorb P, W, G, 750 and J.

**COLUMN** - Chromatographic separation component. A tubular column either filled with a finely divided solid impregnated with a partitioning liquid, or an open tubular column with a liquid film deposited on the walls.

**CONDUCTANCE** - Term used in vacuum technology to indicate the quantity of gas which can pass through an orifice or tube under specific conditions.

**CORRECTED RETENTION TIME, $t(r)$** - Retention time corrected for the pressure drop along the chromatographic column.

**CORRECTION FACTOR** - Syn: RESPONSE FACTOR - Calculation coefficient which corrects the different response of one particular detector to different compounds. Used to convert peak areas to numbers proportional to weight of sample.

**CENTRAL PROCESSING UNIT (CPU)** - The component of a computer system that controls interpretation and execution of instructions.

**DATA BASE** - A collection of data used by a computer program. For GC–MS use, a collection of mass spectra and/or retention times used by computer search systems to identify compounds.

**DEACTIVATED SUPPORT** - Support which has been chemically treated to reduce its surface activity. The most common treatments include acid washing and silanizing with dichlorodimethylsilane.

**DERIVATIVE(S)** - Compound(s) obtained from original sample, usually through chemical reactions, which are more volatile and easily chromatographed. Active protons such as those found in acids, amines, alcohols and phenols are reacted to form more inert esters, ether or silyl derivatives.

**DETECTION LIMIT** - Syn: MINIMUM DETECTABLE QUANTITY - Amount of sample which produces a signal three times the noise level.

**DETECTOR** - Part of the gas chromatograph which constantly monitors the composition of the column effluent by measuring some physical property of the carrier gas and eluted compounds.

**DIATOMACEOUS EARTH** - Syn: KIESELGUHR - Material composed of skeletons of microscopic unicellular algae. Raw material used to make chromosorb(s).

**DIFFERENTIAL FLOW CONTROLLER** - Pneumatic control which maintains a constant flow in spite of the changes in viscosity of the carrier gas due to changes in temperature.

**DIFFUSION PUMP** - Device which produces a high vacuum ($10^{-6}$ Torr) by using the principle of a high speed vapour streaming by a jet.

**DIGITAL-TO-ANALOG (D/A) CONVERTER** - Opposite of analog-to-digital converter.

**DIMETHYLDICHLOROSILANE (DMCS)** - Reagent employed to block the silanol groups of the diatomaceous supports through chemical reaction.

**DIRECT CAPILLARY INTERFACE** - Device that connects the GC to MS by using a capillary tube as a direct connection from GC outlet to MS ion source.

**DISTRIBUTION COEFFICIENT** - Syn: PARTITION COEFFICIENT - Ratio of the concentration of solute or sample in the stationary to the concentration of the same in the gas phase.

**DUAL COLUMN SYSTEMS** - Any chromatographic system in which two columns are used in parallel. The detectors and columns can be identical or different.

**EDDY DIFFUSION** - Phenomenon occurring in packed columns due to lack of homogeneity of packing. Expressed quantitatively by the first term of the Van Deemter equation. Produces peak broadening.

**EFFECTIVE PLATE NUMBER** - Calculated number of theoretical plates employing the adjusted retention time instead of the absolute retention time. Most often used to describe open tubular columns.

**EFFICIENCY OF A COLUMN** - Column characteristic expressed quantitatively by the number of theoretical plates. Efficient columns have many theoretical plates and show only limited band broadening.

**EFFLUENT SPLITTER** - Device which divides the column effluent into two or more streams. Useful when two detectors measure the same sample simultaneously, or when part of the effluent is collected or passed to an auxiliary instrument.

**EIGHT PEAK INDEX** - Mass spectral data compilation which lists the eight most abundant peaks in each mass spectrum. The collection is available in printed form with indexes.

**ELECTRON IONIZATION (EI)** - Method commonly used in a mass spectrometer to ionize sample molecules. Electrons bombard the molecules at low pressures ($10^{-4}$ Torr) to produce fragmentation and ionization.

**ELECTRON MULTIPLIER** - Detection device used in a mass spectrometer in which a series of plates of electron sensitive surface are used to produce a cascade of electrons, as many as $10^6$ for a single charged ion hitting the first plate.

**ELECTRONIC INTEGRATOR** - Instrument which converts the chromatographic signal into a frequency count proportional to peak area.

**ELECTRON CAPTURE DETECTOR (ECD)** - Detecting device based on the decreasing of a background current due to the reaction of sample with electrons produced by emisson of a radioactive source.

**ELUATE** - That which is eluted from the column.

**ELUENT** - General designation of the mobile phase.

**ELUTION ANALYSIS** - Syn: ELUTION TECHNIQUE, ELUTION CHROMATOGRAPHY - The most commonly used technique in chromatography; the sample components are transported by the mobile phase and separated according to their partition coefficients.

**EPA-NIH MASS SPECTRAL DATA COMPILATION** - Joint collection of mass spectra by the U.S. Environmental Protection Agency and U.S. National Institute of Health. Available in printed form or on computer tape or disk.

**EXECUTIVE PROGRAM** - Syn: SUPERVISOR or OPERATING SYSTEM - Master computerprogram that controls the execution of other programs.

**EXTERNAL STANDARD** - see: ABSOLUTE CALIBRATION.

**FERRULE** - Part of fitting (see also FITTINGS).

**FFAP** - Stationary phase specifically developed for fatty acid analysis; consists of Carbowax 20 M transesterified with nitrophthalic acid.

**FILM THICKNESS** - Calculated depth of the layer of stationary phase over the solid support surface or deposited on the inner walls of open tubular columns.

**FITTINGS** - Pieces of plumbing used to connect the column to the instrument. They are commonly nuts and ferrules which make a high pressure, high temperature seal.

**FLAME IONIZATION DETECTOR (FID)** - Detector which measures the changes in conductivity of a hydrogen flame by the introduction of organic vapours.

**FLAME PHOTOMETRIC DETECTOR** - Detector based on the measurement of the emission intensity by a hydrogen flame when sulfur or phosphorous containing compounds are present.

**FLORISIL** - Magnesia silica gel material commonly used for sample clean-up.

**FLOW METER** - Device which allows measurement of the flow rate. Most common is a soap film flow meter.

**FLOW RATE** - Volume of mobile phase per unit time passing through the column, usually reported as ml/min.

**FLOW SPLITTER** - see EFFLUENT SPLITTER.

**FORE FLUSHING** - Technique which groups the early eluting components of a sample (usually onto a second column) and separates them from higher molecular weight components.

**FOREGROUND/BACKGROUND OPERATION** - Syn: MULTITASKING - The ability to process more than one computer program at the same time. One program has highest priority and is executed first (foreground program). When the CPU has to wait for the completion of an instruction, its time can be spent executing instructions from other programs (background programs).

**FORWARD SEARCH** - Usual method of comparing mass spectra during computerized search. Ion abundances in the unknown mass spectrum are compared with ion abundances at corresponding $m/z$ values of each library mass spectrum (see also REVERSE SEARCH).

**FRAGMENTATION** - When a molecule is bombarded with an electron at low pressure, it breaks up into a series of charged fragments. The fragmentation pattern is called a mass spectrum.

**GAS CHROMATOGRAPHY (GC)** - A form of chromatography where the mobile phase is a gas.

**GAS HOLD-UP** - Volume of carrier gas required to transport a compound not retained by the stationary phase throughout the column.

**GAS SOLID CHROMATOGRAPHY (GSC)** - A form of adsorption chromatography where the mobile phase is a gas, and the stationary phase is a solid.

**GAS SYRINGE** - Gas tight syringe. A special syringe, usually 0.1–10 $\mu$l. Equipped with a Teflon plunger which allows gases at high or low pressures to be sampled.

**GHOST PEAK** - Spurious signal due to column or septum bleeding, carrier gas impurities or sample decomposition.

**GOLAY COLUMNS** - Syn: OPEN TUBULAR COLUMNS, WALL COATED OPEN TUBULAR COLUMNS (WCOT), CAPILLARY COLUMNS - Type of column in which the stationary phase forms a very thin film on the inner wall of an open small diameter tubing.

**HEADSPACE ANALYSIS** - Analysis of the vapors above a liquid sample. Commonly used for analysis of foods, flavours, fragrances, etc.

**HEIGHT EQUIVALENT TO A THEORETICAL PLATE (HETP)** - Value obtained by dividing column length with the number of theoretical plates. Taken as indication of column quality.

**HEXAMETHYL DISILAZANE (HMDS)** - Silylating reagent employed to deactivate solid supports by blocking of silanol groups.

**HOT WIRE DETECTOR** - Syn: THERMAL CONDUCTIVITY DETECTOR - The heated component of most thermal conductivity detectors is a spiral of tungsten or a tungsten–rhenium alloy.

**INTERFACE** - (a) GC–MS device used to transfer GC effluent to the MS ion source. It must reduce pressure from atmospheric (760 Torr) to $10^{-5}$ Torr by removing carrier gas while transferring the maximum amount of organic molecules in the GC peak. Types include effusive membrane, jet orifice, and direct capillary. (b) A common boundary between two pieces of hardware (such as CPU and disk drive) or between two systems, which permits transmittal of information between the hardware or systems.

**INTERNAL STANDARD (I.S.)** - Substance used as reference in quantitative analysis. The internal standard is first mixed with standard solutions; later it is added to the unknown, and the ratio of peak heights (or areas) of internal standard and analyte is used for quantitative analysis.

**ION SOURCE** - Part of a mass spectrometer where ions are created at $10^{-5}$ Torr usually by bombardment with 70V electrons.

**IONIZATION DETECTOR(S)** - Detectors based on the production of ionic species in the column effluent; includes FID, EC, HAFID, etc.

**ISOTHERMAL** - Operation of GC column at a constant temperature as opposed to temperature programming.

**JET-ORIFICE INTERFACE** - A device that connects the GC to MS in which the pressure reduction and organic mass transfer occurs by the jet expansion of the GC effluent through an orifice directed into a vacuum.

**KATHAROMETER** - Syn: CATHAROMETER - Thermal conductivity detector.

**KIESELGUHR** - German equivalent of diatomaceous earth.

**KOVATS INDEX (KOVATS RETENTION INDEX)** - Characterization system of the chromatographic behaviour of substances in GC. Normal alkanes are used as reference compounds to establish the reference scale of retention.

**LINEAR GAS VELOCITY** - Rate of carrier gas flow through GC column. It is a variable in the Van Deemter equation.

**LINEAR DYNAMIC RANGE** - decades of concentration within which the detector response is clearly linear.

**LINEARITY** - Proportionality between detector response and amount of sample. A calibration plot with a slope of 1.0 is the ideal case.

**LIQUID CRYSTALS** - Selective stationary phases which have intermediate characteristics between crystalline solids and normal liquids.

**LIQUID LOADING** - Syn: % STATIONARY PHASE - Percentage by weight of stationary phase contained in 100 g of packing material.

**LONGITUDINAL DIFFUSION** - Syn: MOLECULAR DIFFUSION - Second term of the Van Deemter equation which accounts for sample diffusivity and tortuosity of the sample.

**LOWER TEMPERATURE LIMIT** - Minimum temperature at which a stationary phase is useful; below this limit the liquid phase is too viscous to dissolve the sample.

**LOVELOCK DETECTORS** - Syn: ARGON IONIZATION DETECTORS - Several ionization detectors based on the ionization of argon by a radioactive source, and the subsequent ionization of the sample.

**MAGNETIC MASS SPECTROMETER** - A mass spectrometer in which the ion-selection principle depends upon ionic motion in a magnetic field when accelerated in to the field.

**MASS CHROMATOGRAM** - A chromatogram produced by plotting ion intensity of a specific ion vs. spectrum number from mass spectra stored in a computer from a GC–MS run.

**MASS SPECTROMETER** - An instrument which ionizes molecules at $10^{-5}$ Torr pressure and then separates the ions to produce a spectrum of ion abundance versus ion mass. The spectrum is a characteristic pattern for identification of a molecule.

**MASS SPECTRUM** - Characteristic pattern of ion abundance vs. ion mass used to identify an organic molecule.

**MASS TRANSFER** - Band broadening effect due to the lack of equilibrium between the mobile and stationary phase when partitioning the samples; this effect is expressed by the third term of the Van Deemter equation.

**MEAN FREE PATH** - The average distance a molecule will travel before it collides with another molecule. Value depends upon temperature and pressure.

**MEMBRANE INTERFACE** - A device connecting the GC effluent to the MS in which the pressure reduction and organic mass transfer occurs by diffusion of only the organic molecules through a permeable membrane connected to the ion source.

**MESH** - Usual way to characterize the particle size of solid supports. The mesh number indicates the number of wires per inch of the sieve which the particles are passed through.

**METASTABLE ION** - An ion which decomposes into an ion of lower mass and a neutral fragment. Occurs between the ion source and analyzer in a magnetic mass spectrometer.

**METHYL SILICONE(S)** - Most commonly used type of liquid phases. Different degrees of substitution with polar groups produce a wide range of polarities in the stationary phase. SE-30, DC-200, DV-1, DV-101, SP- 2100 are methyl silicone liquid phases.

**MICROPACKED COLUMNS** - Packed columns with an internal diameter of less than 1 mm.

**MICROREACTOR** - Small chamber which contains the reagents to modify the sample before or after the chromatographic separation.

**MIXED PHASE PACKINGS** - Special purpose materials which contain more than one stationary phase.

**MOBILE PHASE** - In GC, the mobile phase is an inert gas such as helium which flows continuously through the column.

**MODEM** - Acronym for MOdulator–DEModulator. An interface device between a computer and a communications link.

**MOLECULAR DIFFUSION** - Effect which produces band broadening expressed by the second term of the Van Deemter equation. The effect is due to the sample diffusion in the gas phase.

**MOLECULAR FLOW** - Describes motion of molecules or ions under low pressure conditions where molecules collide with the walls of the container before other particles. Essentially, all particles move independently of each other.

**MOLECULAR ION** - An ion formed from the sample molecule without fragmentation; gives molecular weight.

**MOLECULAR SIEVE** - Natural or artificial; usually inorganic substances which have well defined properties and are capable of trapping or retaining molecules of specified sizes. An absorbent used in GSC to separate $N_2$, $O_2$, $CH_4$, and CO.

**MASS SPECTRAL SEARCH SYSTEM (MSSS)** - Computerized search system for identifying unknown mass spectra. Several different search options are available.

**MULTIPLE COLUMN SYSTEMS** - Chromatographic techniques in which several columns are employed; in parallel, in series or sequentially for a single analysis.

**NET RETENTION TIME, $t(n)$** - is the adjusted retention time corrected for the pressure drop along the chromatographic column.

**NITROGEN-PHOSPHOROUS DETECTOR (NPD)** - see ALKALI FLAME IONIZATION DETECTOR.

**NOISE** - Random fluctuation of the chromatographic signal. Short term noise (less than 1 s) is often electrical in nature; long term noise can be due to flow rate changes, temperature changes or column bleed.

**NORMALIZATION** - Quantitative method in which the area percent of components from the column is taken as weight percent composition. Of limited usefulness since most detectors give different responses to different types of compounds.

**NORMALIZED MASS SPECTRUM** - A mass spectral plot of ion abundance vs. ion mass in which the most abundant ion is given a value of 100 and other ions ratioed to it.

**OPEN TUBULAR COLUMN** - Syn: CAPILLARY COLUMN, GOLAY COLUMN, WCOT COLUMN, SCOT COLUMN - A column with a hole down the centre, frequently called a capillary column.

**PACKING** - Material contained inside the column responsible for separation.

**PARTITION** - A term describing the distribution of the sample between the mobile phase and the stationary phase.

**PARTITION COEFFICIENT** - Quantitative expression of the partition equilibrium taken as the ratio of concentration of the sample in the stationary phase to the mobile phase.

**PARTITION RATIO (k)** - Syn: CAPACITY RATIO, CAPACITY FACTOR - Column characteristic expressed as the ratio of the retention volumes to the dead volumes.

**PBM SEARCH** - A Probability Based Match computerized mass spectral search system that compares elements of reference spectra with the unknown to aid identification of the unknown.

**PEAK** - The portion of the chromatogram recording the detector response while a single component emerges from the column.

**PEAK AREA** - The area enclosed between the peak (maximum) and the peak base.

**PEAK BASE** - An interpolation of the baseline between the extremities of the peak.

**PEAK HEIGHT** - The distance from the peak maximum to the peak base measured parallel to the ordinate.

**PEAK SYMMETRY** - Characteristic of an ideal Gaussian peak. Lack of symmetry may indicate active sites, poor sample injection technique, too low temperatures, column overloading, incompatible solute and stationary phase.

**PEAK TAILING** - Syn: SKEWED PEAK - Distortion of chromatographic peaks from ideal Gaussian shape seen as an extension of the end of the signal. This may be due to low injection port temperature, active sites in the column, or decomposition of solute.

**PEAK WIDTH** - The segment of the peak base intercepted by tangents to the inflection points on either side of the peak.

**PEAK WIDTH AT HALF HEIGHT** - Same as peak width, but measured at a point one-half the distance from the peak base to the peak maximum.

**PERIPHERAL DEVICE** - Components of a computer system other than the CPU. Examples are terminals, printers, plotters, disk and tape drives.

**PHASE RATIO(B)** - Column characteristic defined as the ratio of mobile phase to stationary phase.

**PLATE HEIGHT** - Syn: HEIGHT EQUIVALENT TO A THEORETICAL PLATE (HETP) - Column length corresponding to a theoretical plate, calculated by dividing the column length by the number of theoretical plates.

**PLATE THEORY** - Theory describing the chromatographic separation as a step by step process in a series of imaginary zones or plates, each one of which results in a distribution equilibrium of the sample between the mobile and the stationary phases.

**POLARITY** - An empirical measure of the separation of charges in a molecule.

**POLYETHYLENE GLYCOL(S)** - Syn: CARBOWAX(CES) - A polar stationary phase used in GLC.

**PORAPAK(S)** - Family of porous polymers (styrenedivinyl benzene) commonly employed in gas-solid chromatography.

**PREPARATIVE CHROMATOGRAPHY** - Chromatographic technique in which the purpose is the separation of sizable amounts of pure materials.

**PRESSURE GRADIENT** - Syn: PRESSURE DROP - Differences between the inlet and outlet pressures in a GC Column.

**PROGRAMMED TEMPERATURE–GAS CHROMATOGRAPHY (PTGC)** - Technique employed to speedup the elution of strongly retained compounds by heating the column according to a chosen temperature/time program as the chromatogram evolves.

**PYROLYSIS** - Analytical technique in which the sample is thermally decomposed prior to the further analysis, e.g. by chromatography. This permits analysis of otherwise difficult samples.

**QUADRUPOLE MASS SPECTROMETER** - A mass spectrometer in which the separation of ions occurs by the action of an electrostatic and radio frequency field created by four parallel rods.

**QUALITATIVE ANALYSIS** - Identification of an organic compound by GC–MS techniques.

**QUANTITATIVE ANALYSIS** - Calculation of sample composition from the data in a GC peak and reference compounds.

**REACTION SITES** - Syn: ACTIVE SITES - Surface points on which the sample can be decomposed; these are usually in the column or on the solid support.

**RECONSTRUCTED GAS CHROMATOGRAM (RGC)** - A chromatographic plot constructed by plotting the total ion current in each spectrum versus the spectrum number.

**RECORDER** - Electromechanical instrument which transforms the chromatographic signal into a graphical record.

**REGISTRY OF MASS SPECTRAL DATA** - Printed collection of mass spectra ordered by molecular weight and elemental composition.

**RELATIVE RESPONSE FACTOR** - see: CORRECTION FACTOR.

**RELATIVE RETENTION TIME** - Syn: SOLVENT EFFICIENCY - Ratio between the net retention time of a substance and that of a standard compound.

**RESOLUTION** - A measure of the separation ability of a GC column or separation of ionic mass by a mass spectrometer.

**RETENTION INDEX** - see: KOVATS INDEX.

**RETENTION TIME (adjusted)** - The retention value obtained by subtracting the time of an unretained component (air) from the uncorrected retention time.

**RETENTION TIME (uncorrected)** - Time elapsed between sample introduction and maximum of response of a GC peak.

**RETENTION VOLUME (corrected)** - Retention volume corrected for pressure drop across the column. Equivalent of adjusted retention time expressed in units of volume.

**RETENTION VOLUME (uncorrected)** - Volume of mobile phase required to elute the peak maximum of a compound. Calculated by the flow rate and uncorrected retention time.

**REVERSE SEARCH** - Method of comparing mass spectra during computerized search. Ion abundances in each library mass spectrum are compared with abundances at corresponding $m/z$ values of the unknown mass spectrum. Ions in the unknown that are not in the library spectrum are not used, therefore multiple co-eluting components can be individually identified.

**SAMPLE LOOP** - A loop of calibrated volume used in a gas sample valve. Normal volumes range from 0.1–10 $\mu$l.

**SAMPLE SPLITTER** - Syn: INLET SPLITTING - Special type of injector which produces two streams of the sample introduced, one is injected and the other is vented to the atmosphere or to a second detector.

**SAMPLE VALVE** - Syn: GAS VALVE - A device used to inject fixed volumes of gaseous sample. May be operated manually or automatically. Provides very reproducible injection volumes. Can be heated but usually only to 200°C.

**SELECTED ION MONITORING (SIM)** - A GC–MS technique in which only a few selected ions are detected and stored as the chromatogram evolves.

**SELECTIVITY** - Measures the selective solubility of a liquid phase for two compounds. High ($\alpha$) values mean high selectivity and easy separation. Expressed as: $\alpha = t(r_1)/t(r_2)$.

**SENSITIVITY** - Term which quantitatively describes the signal obtained per amount of sample introduced.

**SEPARATION FACTOR** - Ratio of retention times of two peaks. Has been replaced by selectivity ($\alpha$) which is the ratio of adjusted retention times.

**SEPTUM** - Self sealing disc through which the sample is introduced into the injection port. Usually a high temperature resistant polymeric material.

**SIGNAL-TO-NOISE (S/N) RATIO** - Ratio of the signal due to analyte response to all other signals (noise). It is generally desirable to maximize the S/N, not just the signal.

**SILICA GEL** - Active solid employed in gas-solid chromatography, chemically can be considered as $SiO_2$.

**SILICONE(S)** - Generic name of a family of numerous stationary phases.

**SILYLATION** - A chemical reaction involving a silane reagent such as trimethyl chlorosilane (TMS). Used to silanize or deactive active protons found in samples on solid support surfaces or on glass tubing.

**SKEWED PEAKS** - Syn: TAILING PEAKS - An assymmetrical peak, not a Gaussian shape.

**SOLID SUPPORT** - Finely divided solid contained in the column whose surface is covered by the stationary phase. Usually a Chromosorb type of material.

**SOLUTE** - Syn: SAMPLE or ANALYTE

**SQUALANE** - Hydrocarbon type of commonly used stationary phase. A $C_{30}$, highly branched saturated paraffin; the reference standard for zero polarity in the Rohrschneider classification system.

**STANDARD ADDITION** - A method of quantitative analysis where standard amounts of sample are added to the unknown. By plotting the increase in peak area or height vs. amount added the concentration in the unknown can be calculated by extrapolation.

**STATIONARY PHASE** - Syn: LIQUID PHASE - Liquid covering the surface of the solid support in packed column or coating the walls of a WCOT column or an active solid contained in the packed column.

**STRIP CHART RECORDER** - Data acquisition system, consisting of a potentiometric measuring circuit and moving chart paper.

**STIRS: SELF TRAINING INTERPRETATION AND RETRIEVAL SYSTEM** - Computerized mass spectral comparison system which can predict the presence of specific substructures in an unknown even if a positive identification is not made.

**THEORETICAL PLATE** - One equilibrium between the mobile and stationary phase. A measure of column efficiency; the more plates, the better is the column efficiency.

**THERMAL CONDUCTIVITY DETECTOR (TCD)** - See KATHAROMETER.

**THERMIONIC EMISSION DETECTOR (TED)** - See ALKALI FLAME IONIZATION DETECTOR.

**THERMISTOR** - Sensing element in some thermal conductivity detectors. Thermistors have a negative coefficient of thermal resistance and generally are most useful at low temperatures.

**TORTUOSITY IN PACKED COLUMNS** - Factor considered in the molecular diffusion term of the Van Deemter equation, expresses the difference between a straight line path and an average real path of a molecule in the chromatographic column.

**TRAPPING** - Condensation of small samples from the column or detector outlet, usually of absorbent materials.

**UPPER TEMPERATURE LIMIT** - Maximum temperature at which stationary phase can be used; beyond this limit there is excessive bleeding or decomposition of the stationary phase.

**VACUUM** - Pressures less than atmospheric created by vacuum pumps.

**VAN DEEMTER EQUATION** - Theoretical relationship which describes sample band broadening as a function of three phenomena: eddy diffusion, gas phase diffusion, and mass transfer.

**WALL COATED OPEN TUBULAR COLUMN (WCOT)** - See CAPILLARY COLUMN, GOLAY COLUMN.

**WATSON BIEMAN INTERFACE** - A concentration device between a GC and a mass spectrometer. The lighter carrier gas is selectively pumped away and the sample becomes more concentrated at a lower pressure.

# SUBJECT INDEX

Printed and bound by CPI Group (UK) Ltd, Croydon, CR0 4YY

03/10/2024

01040428-0015